# Lecture Notes in Networks and Systems

Volume 384

The series "Lecture Notes in Networks and Systems" publishes the latest developments in Networks and Systems—quickly, informally and with high quality. Original research reported in proceedings and post-proceedings represents the core of LNNS.

Volumes published in LNNS embrace all aspects and subfields of, as well as new challenges in, Networks and Systems.

The series contains proceedings and edited volumes in systems and networks, spanning the areas of Cyber-Physical Systems, Autonomous Systems, Sensor Networks, Control Systems, Energy Systems, Automotive Systems, Biological Systems, Vehicular Networking and Connected Vehicles, Aerospace Systems, Automation, Manufacturing, Smart Grids, Nonlinear Systems, Power Systems, Robotics, Social Systems, Economic Systems and other. Of particular value to both the contributors and the readership are the short publication timeframe and the world-wide distribution and exposure which enable both a wide and rapid dissemination of research output.

The series covers the theory, applications, and perspectives on the state of the art and future developments relevant to systems and networks, decision making, control, complex processes and related areas, as embedded in the fields of interdisciplinary and applied sciences, engineering, computer science, physics, economics, social, and life sciences, as well as the paradigms and methodologies behind them.

Indexed by SCOPUS, INSPEC, WTI Frankfurt eG, zbMATH, SCImago.

All books published in the series are submitted for consideration in Web of Science.

For proposals from Asia please contact Aninda Bose (aninda.bose@springer.com).

More information about this series at https://link.springer.com/bookseries/15179

Martha del Pilar Rodríguez García ·
Klender Aimer Cortez Alejandro ·
José M. Merigó · Antonio Terceño-Gómez ·
Maria Teresa Sorrosal Forradellas ·
Janusz Kacprzyk
Editors

# Digital Era and Fuzzy Applications in Management and Economy

Springer

*Editors*
Martha del Pilar Rodríguez García
Nuevo Leon State University
San Nicolás de los Garza, Mexico

Klender Aimer Cortez Alejandro
Nuevo Leon State University
San Nicolás de los Garza, Mexico

José M. Merigó
School of Information, Systems
and Modelling
University of Technology
Sydney, NSW, Australia

Antonio Terceño-Gómez
Department of Business Management
Universitat Rovira i Virgili
Reus, Spain

Maria Teresa Sorrosal Forradellas
Department of Business Management
Universitat Rovira i Virgili
Reus, Spain

Janusz Kacprzyk
Systems Research Institute
Polish Academy of Sciences
Warsaw, Poland

ISSN 2367-3370 ISSN 2367-3389 (electronic)
Lecture Notes in Networks and Systems
ISBN 978-3-030-94484-1 ISBN 978-3-030-94485-8 (eBook)
https://doi.org/10.1007/978-3-030-94485-8



This Springer imprint is published by the registered company Springer Nature Switzerland AG
The registered company address is: Gewerbestrasse 11, 6330 Cham, Switzerland

# Preface

In the face of information technology, digitization, and uncertainty, organizations confront new opportunities and challenges. To take advantage of these opportunities and overcome current and future challenges, it is needed to understand the evolution of these phenomenon. This book "Digital Era and Fuzzy Applications in Management and Economy" convened by the International Association for Fuzzy-Set Management and Economy (SIGEF) jointly with the Universidad Autonoma de Nuevo Leon (Mexico) aims to contribute to the discussion about the implications of fuzzy logic, neural networks, and other intelligent techniques applied to public and private organizations in the current digital era.

Recently, new developments based on socioeconomic and computational changes have shed light on the importance of fuzzy applications in social sciences. The purpose of this publication is to disclose the applications of fuzzy logics and new quantitative methodologies to management and economics in the context of the digital era. This book will be very useful for researchers, practitioners, and graduate students aiming to introduce themselves to the field of quantitative techniques for overcoming digital uncertain environments and developing models to make decisions. It is important to mention that all the chapters were selected from a peer-reviewed process. In this regard, we want to thank all the authors of the chapters, the SIGEF scientific committee, as well as the invaluable contribution of the invited reviewers from different universities who supported the review and selection of the papers received.

This book provides solutions to the organizations problems faced by the lack of information in some cases and the large amount of information. To understand the recent economic and financial behavior, the introduction of the book was carefully written by Dr. Jaime Gil-Aluja. After the 2008 financial crisis, the need for a fast world economic recovery with billions of people connected by mobile devices using the new disrupting technologies, the expansion of internet, and the artificial intelligence was imminent; however, the system was interrupted by a serious external shock: the COVID-19 pandemic, and as Dr. Gil-Aluja mentioned in this chapter "opening up new horizons to which the existing structures need to adapt."

In this scenario, the "Economic humanism self-induced incidences in the circular economy" was introduced.

Then, the following chapters were grouped into four topics of specialized knowledge: finance and economy; management and accounting; methodological issues, and technology and business innovation to provide a scientific approach that contributes to improving the decision-making process in the organizations.

The first part of the book corresponds to the financial and economic analysis. Due to the importance of short-term investing decision, the initial chapter examines the long memory properties in high frequency time series of eight important cryptocurrencies; the results show that high frequency returns exhibit a trend toward a more efficient behavior; on the other hand, high frequency volatility reflects a strong persistence in volatility. Subsequently, in the following chapter, the economic and agronomy risks are studied with a fuzzy decoupled net present value as an improve to the traditional valuation techniques for agricultural commodities. The proposed method allows a more realistic investment estimation in an uncertain environment. As a result of changes in the energy policies in emerging countries, in the next chapter, the performance of energy prices through an elasticity analysis is presented. A quadratic almost demand system estimation indicates that the Mexican demand for energy is inelastic, that is, when prices increase, households do not tend to change their consumption; furthermore, changes in prices have a greater impact on low-income household well-being compared to high-income households.

The second part of the book contemplates the topic of management and accounting. Due to the recently pandemic and the interaction between nurses and patients within an uncertain environment, the need of serious research in the health sector is important. In this logic, the first chapter of this section proposed the use of fuzzy qualitative comparative analysis as an effective method for understanding the cognitive heuristic decision-making in the emergency department of a private–public hospital in Naples; the authors also proposed some practical implications to improve the triage process. On the other hand, the next chapters are oriented to assess accounting issues. To analyzes whether ESG risk information disclosures impact on earnings per share to confirm accounting conservatism in Mexico, a multifactor fuzzy regression is proposed. The results suggest the presence of accounting conservatism in the Mexican capital market in 2020 due to the negative disclosures of ESG risks since the "bad socially responsible news" increases the impact of the market reaction to financial performance. Finally, since companies need to evolve toward the new digital era as a result of the technological advances in the government taxation mechanisms, a research to assess the effect of the digital taxation mechanisms in Mexico is presented. The results show that the implementation of digital taxation directly benefits income tax collection.

The third block of manuscripts is related to methodological issues in order to analyzes uncertain environments. First, a methodological proposal is presented to analyze the poverty through educational lag from a complex network and optimization approach. Then, using a spatial analysis, the following chapter explores the existence or lack of a process of convergence among the economic production of Mexican firms in different cities; the results show that the geographic location determines the successful of the firms. A final method in this section is proposed to asset the cryptocurrencies investment industry. The on-chain data that uses metrics from the blockchain of the underlying asset and technical analysis is applied to predict investor sentiment for the cryptocurrency markets.

Finally, settled on the Fourth Industrial Revolution, three chapters are presented in the technology and business innovation section of the book. The first chapter provides some insight into designing a visual graph-shaped frontend for two flagship deep learning software platforms. The proposed frontend, called Visual Keras&Autokeras, is attempting to visually emulate all the APIs related to Keras Functional model and AutoKeras AutoModel in a codeless environment at any level of complexity. On the other hand, as price optimization is an important research topic for business and economy, the next paper proposes a novel approach based on two major components: the predictive component using parametric and non-parametric machine learning techniques, and the forecast component processed into an optimization model to analyze cross price elasticities in order to maximize the revenue for a retailer while keeping control of traffic and assortment at the stores. Finance is not the exception in these digital changes; in this sense, a fintech analysis is presented at the end of the book. The objective of the final chapter is to compare an investment portfolio that utilizes similar strategies to those of an Robo Advisor against an investment portfolio that makes decisions through a consensus of valuation analysts. To compare both portfolios, a Fuzzy Jensen's alpha is estimated. The results indicate that both strategies succeeded in surpassing the benchmark, but the analysts' portfolio has accelerated its growth since 2018 against the Robo Advisor portfolio. However, the Robo Advisors' portfolio has a higher possibility of obtaining abnormal or unexpected returns than the analyst's value investing portfolio, given the systematic risk involved.

As we have mentioned, the treatment of uncertainty in the economic and business analysis is fundamental and requires methods compatible with the uncertain environment in the recent digital era since most of the traditional models have been overtaken by this reality when trying to make decisions with uncertain information in a big data word. In this sense, fuzzy logic, optimization approaches, complexity science, economic process modeling under uncertainty and geographic context, fintech analysis, deep learning and other artificial intelligent techniques applied to different organizations can be powerful tools in the decision-making process. The selected and peer-reviewed chapters in this book invite to delve into these issues with real applications that can undoubtedly be of great utility to both researchers and practitioners.

Martha del Pilar Rodríguez García
Klender Aimer Cortez Alejandro
José M. Merigó
Antonio Terceño-Gómez
Maria Teresa Sorrosal Forradellas
Janusz Kacprzyk

# Organization

The International Association for Fuzzy-Set Management and Economy (SIGEF) joint with reserachers in the Universidad Autónoma de Nuevo León (UANL) prepared the blind arbitration process for the Springer book series with the title "Digital Era and Fuzzy Applications in Management and Economy" and book series on "Advances in Intelligence Systems and Computing."

Academic comittee is working to arrange the publication of the best papers and chapters. Manuscripts will be selected for potential publication in special issues or Springer book. Twenty-nine researches have been submitted, and 13 chapters were selected for publishing in Springer book considerng the arbitraje doble blind. The book has the participation of various colleagues from different countries all over the world, such as Argentina, Croatia, Spain, Italy, Perú, and México.

## Guest Editors

Martha del Pilar Rodríguez García
Klender Aimer Cortez Alejandro
José M. Merigó
Antonio Terceño-Gómez
Maria Teresa Sorrosal Forradellas
Janusz Kacprzyk

# Contents

## Methodological Issues



## Technology and Business Innovation

# Economic Humanism Self-induced Incidences in the Circular Economy

Jaime Gil Aluja(✉)

Real Academia de Ciencias Económicas y Financieras, Via Laietana, 32, 08003 Barcelona, Spain
secretaria@racef.es

**Abstract.** The Circular Economy has proven to be one of the main advances in trying to solve the serious problem of environmental degradation. In this work we present an algorithm, based on the reticular theory, to detect the incidence of actions (incident elements) linked to the circular economy on products (incident elements). For this we establish both the direct incidents, which are easier to determine by the experts, and the self-induced incidents whose determination is not so simple. Starting from these incidents, the semi-accumulated flows are established at different stages of the algorithm, until the final accumulated flows are obtained. These contain all possible incidents, both direct and indirect, obtaining a network of incidents that will allow determining the best policies to obtain the desired result. We present the application to a specific case, although the significant contribution is the methodological proposal that we understand in accordance with human thought and that allows those incidents that are not direct and therefore difficult to detect by the human mind to appear.

**Keywords:** Circular economy · Reticular theory · Incidences algorithm

## 1 External Impacts in an Evaluative System

Information, realities, actions, reactions, strategies, but basically and sometimes as protective umbrellas: politics. A subtle word, so often prostituted, so often maltreated. And a waste collecting recipient: the economy.

Not so long ago, in 2008, a serious internal crisis was generated and exploded as a result of the financial "imbalances" that invaded practically every area of the economic systems.

Some short-term measures were taken to repair the damage caused and some medium and long-term ones outlines to adapt to the new realities of our world, which had been shaken up as much by internal impacts on the economic system as external ones, caused by the "toxic" activities of the leaders of life in society (public, institutional, and business managers and, of course, economic units of consumption and investment).

Having overcome the financial crisis, the outlook for recovery and economic expansion was looking increasingly bright and imminent when the system was interrupted by a serious external shock in the form of the Covid-19 pandemic which, driven by the confusion into which the health, political, and economic organization had been thrust,


M. d. P. Rodríguez García et al. (Eds.): XX SIGEF 2021, LNNS 384, pp. 1–21, 2022.
https://doi.org/10.1007/978-3-030-94485-8_1

changed the existing "normality" with its accepted routines, opening up new horizons to which the existing structures need to adapt.

It is not an easy task. The human mind is slow to account for changes that occur, but at least the economic structures appear not to have suffered any internal dam-age. Recovery is happening fast.

The fact is that nowadays, the events that have varying degrees of impact on economic activity happen sporadically, but continuously. The shocks suffered by the economic system are no longer separated by 10-year gaps, but by just a few months, with the subsequent need for repair and adaptation to the new scenarios.

Just a few short weeks ago, a new Palestinian-Israeli conflict broke out, with missiles, death, and devastation. At the time of writing these words, sorrowful scenes of minors trying to reach Spain from Morocco via Ceuta are being played out in basic seafaring craft, swimming and scaling fences, sometimes straight to their death. We are the stunned observers of painful and emotive scenes of "illegal" children being embraced by Spanish soldiers.

The speed with which these changes are happening leads us to a first conclusion: any adaptations must be flexible so that they can be swiftly modified, and without involving huge costs and the need for great effort.

Meanwhile, the economic studies and research undertaken to underplay the negative incidence of economic cycles and to highlight the opportunities their emergence always offers, continue to be produced with the support of principles, techniques, and operators of the strictest mechanical philosophy. Nonetheless, economic science has not changed since its birth. Its models, methods, and calculation procedures are little effective or entirely ineffective when dealing with the complex problems and uncertainties of our current realities.

It is true that from time to time some new concepts, processes, or techniques that could potentially open new paths come to the attention of economic research teams. This is the case of the appearance of the "Fuzzy Sets" work of Lofti Zadeh [1], which we proudly contributed to.

While different in terms of its potential and originality, the concept of the "Circular Economy" is another such example. In essence, it consists in reintroducing an old method of economic management adopted during and after the Spanish Civil War, reborn now at odds with the waste involved in the consumption and production of goods and the investment that a large part of our societies take part in in an UNSHARED prosperity.

And what is more, this distinct lack of sharing not only occurs between territories but is also often missing in the heart of a territory.

The recent images of undernourished and hungry children roaming around Ceuta, and the sadly recurrent images of young and old alike sleeping outdoors in the cold of winter and rummaging in our city's rubbish bins looking for what others have thrown away to fill their stomachs must prompt consciences to not remain indifferent to such serious imbalances. It is now appropriate to pose the following question:

What attitude must we take in the face of this desolate scenario?

It is not the investigative sector's responsibility to carry out activities that belong in the political or legal realm, but it is their responsibility to work to supply these sectors with the elements of rigor and effectiveness to underpin any valid decisions they make based on legality and ethics.

We have been doing it this way up till now and our proposal is to do it the same way in the future.

And not only have we done it this way, but we offer our findings with no compensation whatsoever and with total freedom of use.

## 2 A Road to the Humanist Economy

From an operative point of view, we must highlight the construction of humanist algorithms designed to find effective solutions to serious and complex problems.

Among the latest humanist algorithms created, published, and offered we can find the "Portugal Algorithm" [2] designed to enhance cohesion between the countries and zones of the Iberian Peninsula.

The "Algorithm for allocating immigrants" [3], which seeks compatibility between immigrants' possibilities and aspirations and the desires and needs of the welcome companies and institutions, is another of them.

The purpose of a variant of this algorithm is to solve the enormous problem of the MENAS (non-accompanied minors) to better allocate immigrant minors to welcome families, taking into account the human, social and educational characteristics of the first and the desires and possibilities of the second. We believe we do not have to spell out the high degree of subjectivity involved in some of the criteria set out to guide this allocation.

Nowadays, with the serious situation in Ceuta, the "Algorithm to allocate non-accompanied minors" fits perfectly with the problem that has now emerged. And even though political agreements are periodically made, we are seriously concerned that this problem will persist for many years to come. We reiterate our willingness to help resolve the situation.

The "Algorithm for inter-generational harmony" [4]. This main aim of this algorithm, which is generalizable to harmony in all aspects of life in society, is to make compatible the tasks involved in a workplace such that that they complement the work destined for various generations of humans.

As we speak, we are working with Dr Jean-Jacques Askenasy, Professor of Neurology of Tel-Aviv University on an algorithm for the early detection of neurological disorders.

These and other works have a humanism that impregnates as much the tools used to resolve the problems in hand as the problem itself.

The general aim to some degree is to find a way to prevent problems from happening before they do and to make sure they do not appear later, manifesting themselves in all their harshness.

The creation of these and other algorithms marks the end of our venture into humanist economics in which, in addition to the rational components, subjectivity and emotional aspects explicitly intervene in the formalization of human decisions.

Looking back over our research, we realized that in a some of our works we obtained acceptable results without the need to lead the way with numbers, in the sense that we could evade the operators belonging to arithmetic.

Therefore, the sum could be substituted by an addition, the difference by a distance, the derivatives to optimize by the maximization of minimums, and so on. We had already created a humble but solid body of doctrine, ranging from the principles to techniques for its immediate application.

Four concepts with their operators were sufficient to find suitable solutions for any economic problem in a context of uncertainty: relation, consolidation, allocation and ordering [5]. In 1999, I published a work in Kluwer [5], which covered my research work developed from the four theories we formalized.

"Numerical mathematics" of uncertainty and "non-numerical mathematics" alone already provided a body of technical knowledge capable of dealing with economic phenomenology with humanist tools.

In its widest sense, this has only been possible by anticipating what society will be like wherein human relations with other humans will take place. We had made previous attempts to do this in view of the advances in transhumanism and dataism, along with the risks and important advantages and opportunities these lend, and to which we were among the first to contribute [6].

During times of profound and rapid change, and on the difficult-to-predict courses on which the world is nowadays set, we need some rays of sunshine to avoid the pitfalls of delirium, disorientation, and disillusion, the vitamin supplements for dehumanization.

To confront these dangers, there is also a serious need to strengthen cooperation in all areas of our shared existence, but particularly in the field of scientific investigation. To weaken solidarity is to erode humanism. To convert solidarity into selfishness is to reduce it to a mechanism: using machines to serve humankind is a battle we can win.

We continue to be committed to placing all the effort we are capable of at the service of idea of a human future, generator and driver if the mechanical at its service.

## 2.1 Circular Economy: From a Past Inheritance to a Future Opportunity

Meanwhile, these and other actions, from thousands of millions of individual acts to the consolidated habits of companies, corporations, and institutions, are causing further deterioration of the planet. Slowing this down at the very least, and if possible changing the trend, requires the tenacity of everybody, so that the sum of many "not much in the way of positives" can be converted into an interesting collaborative cooperation.

At this point, we would ask your permission to leave aside the air of formality for a moment to examine some of the small realities of our past and present.

Some years ago, in the last decades of the XX century, and like with other facts and phenomena, concepts and processes, an attractive name was coined to designate a way to carry out economic activity: circular economy.

This term somehow managed to encompass activities in the form in which they took place in our country in times of shortage, and are still taking place nowadays in economically deprived communities. What do realities like the second-hand market, re-used textbooks, second-hand bike sales, and hand-me-down clothes say to us, and especially to us older members of society?

Words that used to have no pejorative connotations at all, like junk dealer, rag and bone man (the typical French "chiffonier"), and scrap dealer, have also stuck in our memory. Does anybody remember going to buy food and drink in glass bottles or containers and taking them back once they were empty? Well, today's circular economy has something of all this in it, but back then it was physically dispersed and formally unconnected. Its only reason to be was survival.

And then low cost, easily manipulable and transformable products and raw materials appeared, and the "throwaway society" was born. Some jobs disappeared and others appeared: a simple fact of life.

But lands and seas have filled up with rubbish, pollution in our villages and cities has increased, and we are now suffering the consequences of the warming of our planet.

Actions with varying degrees of effectiveness and efficiency seek to halt this destructive process, implementing measures designed not only to limit the existing accelerated squandering but also slow down the soiling of both our planet and the part of the Cosmos nearest to us.

We may think ourselves and our humble occupations more important than they really are in terms of this crusade to safeguard our future. But what is intolerable is sitting back and watching how we are jeopardizing the progress of future generations.

To this effect, we simply claim that it is within our scope to help improve the structures and functioning of the circular economy.

And the way we think we can do this is by completing its conceptualization, specifying its operation, and providing it with operational techniques using artificial intelligence.

In studies related to the circular economy, the lifecycle of a product is usually described as a series of stages that start with raw materials entering a process of production, manufacture, and use, representing the end of the product's first life, and its eventual utilization (with or without transformation), thus closing the circle with a new use.

We can of course imagine all the variants. However, for our purposes, this simple but interesting description is enough to understand that this process can be expressed through a flow in a network, which can be as simple or as complex as required.

This way, we can largely take advantage of the schemes and operators of the reticular theory, which we have so often used to resolve other problems [7], although to do so it is recommendable to somehow define some aspects that represent the circulation of the flows in the networks. This is what we are now going to do.

However, what cannot be forgotten is the fact that there are other circuits that are just as important and without which we cannot talk about the circular economy. We are talking about the ones that come to rest where the flows in the circle start.

We will try to fill this potential gap with what we believe will help open up a new path to taking good decisions.

## 3 Incorporating the Notion of Incidence into the Circular Economy

To do so, let us go back to the notion of incidence. Incidence is usually associated somehow with a relationship, in the same way as is causality (cause and effect relationship). The "causes", incident elements, come together and jointly intervene to some degree or level in the "effects", another set of elements impacted on.

By way of illustrating our proposal, we are going to introduce a formal process that starts with establishing two sets of reference A and B, comprised of the incident elements $a_i$, I = 1, 2, …, n, and the elements impacted on $b_j$, j = 1, 2,…, m.

As an example, and only an example, potentially included among the incident elements are government measures, press articles and releases, waste collection points in cities, intermediary companies such as Vinted, Wallapop, and Poshmark, home collection services, recovery solidarity organizations such as Humana, Arrels, Caritas, the Red Cross, etc., and commercial second-hand companies such as Tuvalum, Thingeer, and Percentil, etc.

Again, just as a reference, other incident elements could include the inventories of products and goods in storage for their re-incorporation in the circular economy: individual transport vehicles such as bicycles and skates, etc., furniture and other household items, clothing and accessories, glass containers, books and school equipment, and food leftovers.

Once the elements that make up the incident sets $A = \{a_1, a_2,\ldots, a_i, \ldots a_n\}$ and $B = \{b_1, b_2, \ldots, b_j, \ldots, b_m\}$, the basic concepts are established and the operators required for their later use are chosen.

We propose the following, produced for the research work we presented at the 6th International Conference: "Economics Scientific Research. Theoretical, Empirical and Practical Approaches" on 10–11 October 2019 in Bucharest [8]:

1. Channels of incidence: arcs of the networks and subnets through which the incidence flows.
2. Incidence flow: the level of flow through the channels. It is evaluated in the interval [0; 1]
3. Incidence deposits: arc peaks in a network where the incidence flow can reach or leave
4. Channel confluence: the peaks of a network can be deposited where the incidences that arrive from two or more channels flow.
5. Incidence distribution center: peaks of a network where two or more channels exit.
6. Dilution of incidence flows: a phenomenon that occurs in an incidence deposit when the arrival flow is less or greater than that generated by the deposit itself. The deposited flow is reduced to the lower of the two.
7. Incidence Flow Evaporation Threshold: The degree or level below which the incidence flow stops in a reservoir and stops passing through the next channel or channels.

By establishing these concepts, we can present the most important aspects we propose to achieve the following two objectives:

I. Obtain the total number of incidences from the incident entities to the affected entities, and detect the incidents that act as the most important intermediaries.
II. Establish the level or degree of incidence flows that reach the final deposits.

Within the framework of a conference of the type that would be interested in accepting this research work, we believe it is necessary to insist that the incidence evaluations take place using the endecadarian system in the confidence interval [0, 1].

Despite this, and to provide a solid basis for our previous explanation, we are now going to replicate the basic principles of the theory of incidences:

a. The natural quality of the human being doesn't allow the use of just the two positions of the binary system: the incidence exists or it doesn't exist. Let's introduce nuance by appealing to multivalent logic and, mathematically, to fuzzy sets.
b. Incidences propagate with a chain of relations through a network. To formalize this, we use the image of a flow of incidences with a degree or level of intensity that flow through the connected channels with a capacity degree or level.
c. Incidence is a notion with a highly subjective content which, consequently, cannot be generally treated through the operators with which determinism, certainty and hazard are concerned, as it is not the object of quantification (objective numerical assignment) but of an assessment (subjective numerical assignment). Determinism is not a good bet for human adventure. Everything evolves, and adaptability has become essential.
d. If, in Boolean algebra, values are taken from the set {0; 1}, in multivalence we establish that these are included in the interval [0, 1]. In this regard, we propose an endecadarian numerical-semantical correspondence as an indication (we can use another one too) [8].

## 4 Direct and Self Induced Incidences

Having set out the most essential concepts and basic principles for developing our work, we are now in a position to start the stages needed to incorporate the theoretical and technical elements of the self-induced incidences into the study of the circular economy.

First, we will proceed in obtaining the direct incidences, or in other words those that represent the "degree" or "level" of incidence of all $a_i$/i = 1, 2,…n, on all $b_j$/j = 1, 2, …, m, without incidences that act as intermediaries.

For greater clarify, we will represent the grade or level of incidence of all ai/i = 1, 2, …n on all $b_j$/j = 1, 2, …, m, by means of $(x_i, y_j)$/i = 1, 2, … n; j = 1, 2, …, m. The entire valuation $(x_i, y_j)$/i = 1, 2, … n; j = 1, 2, …, m is included in [0, 1].

The set of valuations in [0, 1] of the direct incidences of $a_i$/i = 1, 2, …n on $b_j$/j = 1, 2, …, m, can be represented by a fuzzy matrix $\left[\underset{\sim}{M}\right]$ like the following one (Fig. 1).

Where the valuations of all the pairs ($a_i$, $b_j$), or in other words $(x_i, y_j)$, i = 1, 2, … n; j = 1, 2, …, m, are included in [0, 1].

$$\forall(a_i, b_j) \in \left[\underset{\sim}{M}\right] \tag{1}$$

$$(x_i, y_j) \in [0, 1] \tag{2}$$

$[\underset{\sim}{M}] =$

| B / A | $b_1$ | $b_2$ | | $b_m$ |
|---|---|---|---|---|
| $a_1$ | $(x_1, y_1)$ | $(x_1, y_2)$ | | $(x_1, y_m)$ |
| $a_2$ | $(x_2, y_1)$ | $(x_2, y_2)$ | | $(x_2, y_m)$ |
| | | | | |
| $a_n$ | $(x_n, y_1)$ | $(x_n, y_2)$ | | $(x_n, y_m)$ |

**Fig. 1.** Direct incidences matrix.

The matrix $\left[\underset{\sim}{M}\right]$ therefore expresses the direct incidences of all the elements $a_i/i = 1, 2, \ldots n$, the set of incidences of the set A, on all the elements of $b_j/j = 1, 2, \ldots, m$, the set of affected elements of the set B.

We will now formally express the self-induction of incidences, although it may be pertinent to first clarify that what is understood as self-induction of incidences are the relations of incidences that take place among elements in the same set, or in other words, among the elements in A and, likewise, among the elements in B, and including the incidence of an element with itself.

To this effect, it is considered that, given a referential A of incident elements on the referential B of affected elements, there are also some incidences of $a_i/i = 1, 2, \ldots n$, on $a_h$, $h = 1, 2, \ldots, n$, whose corresponding valuations $(x_i, x_h) \in [0, 1]$, i,,h = 1, 2, …, n, form a fuzzy matrix $\left[\underset{\sim}{A}\right]$, which is square and reflexive (Fig. 2).

$[\underset{\sim}{A}] =$

| A / A | $a_1$ | $a_2$ | | $a_n$ |
|---|---|---|---|---|
| $a_1$ | $(x_1, x_1)$ | $(x_1, x_2)$ | | $(x_1, x_n)$ |
| $a_2$ | $(x_2, x_1)$ | $(x_2, x_2)$ | | $(x_2, x_n)$ |
| | | | | |
| $a_n$ | $(x_n, x_1)$ | $(x_n, x_2)$ | | $(x_n, x_n)$ |

**Fig. 2.** Self-induced incidences matrix of incident elements.

where $(x_i, x_i)$, i = 1, 2, …, n, are equal to the unit.

Similarly, given a referential of affected elements, B, of another referential of incidences A, these are also incidences of their own elements. Therefore, there are some incidences of $b_j/j = 1, 2, \ldots, m$, on $b_k/k = 1, 2, \ldots, m$, and their valuations.

$(y_j, y_k) \in [0, 1]$, j, k = 1, 2… m form a square and reflexive matrix, $\left[\underset{\sim}{B}\right]$ (Fig. 3).

In the language of flows in networks we have defined, we can say that between two sets, one made up of primarily incident elements, A, and the other made up primarily of affected elements, B, there is also a flow of incidences of A to B, whose direct degree or level, the volume, is represented by a fuzzy matrix $\left[\underset{\sim}{M}\right]$, and two other flows of incidences that take place from the elements in A on the elements in A, and including the incidences of an element on itself and other flows of incidences from the elements in B on the elements in B, also including the incidences of an element on itself.

The fuzzy matrices $\left[\underset{\sim}{A}\right]$ and $\left[\underset{\sim}{B}\right]$, express the valuations of the self-induced incidences.

To this effect, we now have valuable information contained in the three incidence networks: the network of direct incidences and the two networks of self-induced incidences.

$\left[\underset{\sim}{B}\right] =$

| B \ B | $b_1$ | $b_2$ | … | $b_m$ |
|---|---|---|---|---|
| $b_1$ | $(y_1, y_1)$ | $(y_1, y_2)$ | | $(y_1, y_m)$ |
| $b_2$ | $(y_2, y_1)$ | $(y_2, y_2)$ | | $(y_2, y_m)$ |
| ⋮ | | | | |
| $b_m$ | $(y_m, y_1)$ | $(y_m, y_2)$ | | $(y_m, y_m)$ |

**Fig. 3.** Self-induced incidences matrix of affected elements.

where $(y_j, y_j)$, j = 1, 2, …, m, are equal to the unit.

## 4.1 Determining the Total Level of Flow of the Incidences

We have now reached a point in our work where we assume that the degree or level of the flows of direct incidences of set A of primary incidences on set B also of primary incidences, without considering the possible incidences through elements that act as intermediaries, are known.

The question begged is do the total direct incidences always have a higher degree or level than the other totals if they take place through other intermediary incidences? An affirmative response is impossible given that with the intermediary/intermediaries a flow or flows can accumulate, brought about by the action of the intermediaries.

As is usual in our research, for the potential addition of flows, we have chosen the max-min convolution operator.

Given that the matrices that express the degree or level of self-induced incidences $\left[\underset{\sim}{A}\right]$ and $\left[\underset{\sim}{B}\right]$ are square and reflexive matrices, we can use the associative property but not, and this is important to remember, the non-commutative property.

If we designate as $\left[\underset{\sim}{M}^*\right]$ the matrix that expresses the global degree or level of the total flow of incidences, we get:

$$\left[\underset{\sim}{M}^*\right] = \left[\underset{\sim}{A}\right] \circ \left[\underset{\sim}{M}\right] \circ \left[\underset{\sim}{B}\right] \quad (3)$$

Where:

$$\left[\underset{\sim}{M}^*\right] \underset{\sim}{\supseteq} \left[\underset{\sim}{M}\right] \quad (4)$$

The procedure for the accumulation of flows is very simple. It is sufficient to first find the degree or level of the semi-accumulated flows of incidences with the max-min convolution $\left[\underset{\sim}{A}\right] \circ \left[\underset{\sim}{M}\right]$ for each element $(a_i, b_j)$, i = 1, 2, …n; j = 1, 2, …, m, making:

$$\left(x'_i, y'_j\right) = \bigvee_j \left((x_i, x_h) \bigwedge (x_h, y_j)\right) \quad (5)$$

i, h = 1, 2, …n.
j = 1, 2, …, m

The result $\left[\underset{\sim}{A}\right] \circ \left[\underset{\sim}{M}\right]$ includes the possible addition of the direct incidences and the self-induced incidences of the elements that primarily exercise the role of direct incidences on the elements of the set primarily made up of affected elements of set B. The following is satisfied:

$$\left[\underset{\sim}{A}\right] \circ \left[\underset{\sim}{M}\right] \underset{\sim}{\supseteq} \left[\underset{\sim}{M}\right] \quad (6)$$

We will now proceed to accumulate the degree or level of the flow of incidences corresponding to the self-induced incidences of set B.

For each element, $(a_i, b_j)$, i = 1, 2, …n; j = 1, 2, …, m:

$$(x_i{}^*, y_j{}^*) = \bigvee_j \left(\left(x'_i, y'_j\right) \bigwedge (y_j, y_k)\right) \quad (7)$$

i = 1, 2, …n
j, k = 1, 2, …, m:

The result $\left[\underset{\sim}{M}^*\right] = \left(\left[\underset{\sim}{A}\right] \circ \left[\underset{\sim}{M}\right]\right) \circ \left[\underset{\sim}{B}\right]$ contains the addition of the levels of direct incidence and those of self-induced incidence for set A of primarily incident elements and set B of primarily affected elements: the picture is now complete.

$$\left[\underset{\sim}{M}^*\right] = \left[\underset{\sim}{A}\right] \circ \left[\underset{\sim}{M}\right] \circ \left[\underset{\sim}{B}\right] \tag{8}$$

We now have a flow of network incidences made up of all the possible flows, without error or omission, which constitutes the path with optimal flow, if the decisive criteria of prudence, represented in our case by the max-min convolution operator, is accepted.

With huge and merited pride and satisfaction we would like to express our recognition of the work of Kaufmann and Gil Aluja: "Models for the research of forgotten effects" [9], published in 1988, immediate predecessor in many ways of this work to which we feel enormously indebted.

## 4.2 The Complexity of Incidence Flows in the Circular Economy

Throughout this work, we paid special attention to two interdependent stages of the circulatory process: on the entry of "things" into the circulatory system, incident elements, and the new use of "things", exit from the circulatory system, affected elements.

For the system to work, and to able to suitably regulate its activity in line with the needs of increasing or decreasing the volume of entries into/exits from the circle, we propose an algorithm that establishes which element of set A, the incident elements, is needed to achieve the optimum volume of exits of an element of set B, the affected elements.

To this end, we have chosen a formula that has produced good results for us in other works: the construction of the stages of an algorithm with the information obtained in an experimental stage. To do so, the numerical data obtained must be considered the result of a mathematical conjecture.

We will begin with the presentation of set A of incident elements:

$a_1$ = government regulations and measures.
$a_2$ = press articles and advertising.
$a_3$ = waste containers in villages and cities.
$a_4$ = home collection services.
$a_5$ = recovery solidary organizations (Humana, Arrrels, Cáritas, the Red Cross, etc.)
$a_6$ = intermediary companies (Vinted, Wallapop, Poshmark, etc.)
$a_7$ = commercial second-hand companies (Tuvalum, Thingeer, Percen-til, etc.)
$a_8$ = illegal street selling.

Of course, this is not an exhaustive list but just a small sample we believe to be sufficient for a research work of this type.

We will now number the elements we are going to consider represent the products, goods, and objects that flow in the circulatory process and that make up set B de affected elements:

$b_1$ = cars
$b_2$ = individual modes of transport (bicycles, skates, etc.)
$b_3$ = furniture
$b_4$ = household items
$b_5$ = household appliances
$b_6$ = plastic products
$b_7$ = glass packaging
$b_8$ = clothes
$b_9$ = office equipment
$b_{10}$ = schoolbooks and equipment
$b_{11}$ = gift and decorative items
$b_{12}$ = leftover food.

It is, therefore, an approach that prioritizes the novel aspect of incorporating some theoretical and technical elements into a way of running the economy, which was not previously mainstream due to its secondary effects, unlike those of the business economy. It is not enough to simply place and replace products in the market, with profits. The circular economy also seeks, as a matter of priority, to provide a community service: to give food to the needy, shelter to the homeless, and clothes to those with none, and to conserve the planet's atmosphere, clear lands and seas of waste, combat global warming, and so on.

To do so, we have the pertinent concepts and operators, and the technical operatives belong to the same group as those used below.

We have reduced to the minimum, eight and ten, respectively, the elements to be used from the set of incident elements and from the set of affected elements. We believe these are sufficient numbers to express the relations of incidence taken two by two using a rectangular 8 × 12 fuzzy matrix $\left[\underset{\sim}{M}\right]$. The valuations made by a committee of experts on the degree or level of incidence of each element of the set of incident elements A on each element of the set of affected elements B, expressed using the endecadary system in [0, 1], are placed in the corresponding boxes.

To also take into account the self-induced incidences, we are now going to construct two more matrices, which will of course be square and reflexive. In the first, we will place the valuations of the incidences of each incident element with the other incident elements, including the incidence with itself (the total), and in the second we will place the valuations of the incidences of each affected element with the other affected elements, including the incidence with itself (also the total).

In a small 8 × 12 matrix like the one proposed, if we take into account a single incident element and a single affected element, the potentially existing flows can be represented in a network like the ones we have previously presented, as shown below in Fig. 4 [9].

To take into account the large number of channels through which the incidence flows of the 8 elements of set A of incident elements towards the 12 elements of set B of affected elements, it is sufficient to join each of the $a_i/i = 1, 2, \ldots 8$ with each of the $b_j/j = 1, 2, \ldots, 12$, which would entail forming 8 × 12 = 96 times the previous figure.

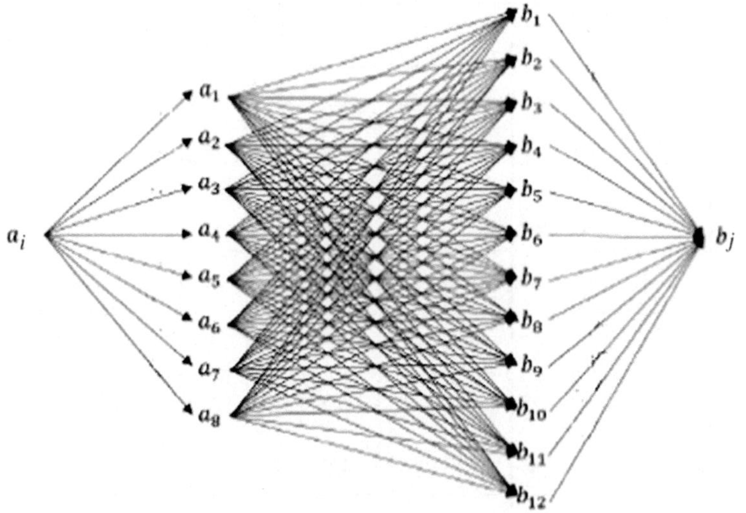


**Fig. 4.** Flow network between an incident element and an induced element.

### 4.3 Algorithm to Optimize the Flows in the Circular Economy

With the aim of avoiding the unnecessary risk of descending into confusion, we are going to number the stages of the algorithm we are proposing, together with the valuations or measures established by the experts, and indicating the operators that will be used in each stage.

1. Formation of set A of elements that act as incident elements: $A = \{a_1, a_2, \ldots a_8\}$
2. Formation of set B of elements that act as affected elements: $B = \{b_1, b_2, \ldots, b_{12}\}$
3. Construction of the matrix $\left[\underset{\sim}{M}\right]$ of valuations $(x_i, y_j)$, $i = 1,2,\ldots 8; j = 1, 2, \ldots 12$.

This stage consists only in collecting the information provided by the committee of experts and its placement in the corresponding boxes in the matrix.

Once these tasks have been completed, we have the fuzzy matrix $\left[\underset{\sim}{M}\right]$ of primary incidences and, therefore, of the degrees and levels of the primary flows that circulate in the network (Fig. 5).

The incidences of all the incident elements $a_i/i = 1, 2, \ldots 8$, on all the affected elements $b_j/j = 1, 2, \ldots, 12$, gathered in the matrix $\left[\underset{\sim}{M}\right]$ show the simplicity of the network of flows in comparison to the one that is produced when the self-induced incidences are incorporated.

$[\underset{\sim}{M}] =$

| $\underset{\sim}{M}$ | $b_1$ | $b_2$ | $b_3$ | $b_4$ | $b_5$ | $b_6$ | $b_7$ | $b_8$ | $b_9$ | $b_{10}$ | $b_{11}$ | $b_{12}$ |
|---|---|---|---|---|---|---|---|---|---|---|---|---|
| $a_1$ | 0.6 | 0.8 | 0.4 | 0.3 | 0.6 | 0.9 | 0.8 | 0.2 | 0.4 | 0.7 | 0.2 | 0.7 |
| $a_2$ | 0.5 | 0.7 | 0.2 | 0.1 | 0.4 | 0.8 | 0.7 | 0.5 | 0.5 | 0.6 | 0.2 | 0.8 |
| $a_3$ | 0 | 0.2 | 0.8 | 1 | 0.9 | 0.8 | 0.6 | 0.7 | 0.9 | 0.7 | 0.6 | 0.7 |
| $a_4$ | 0.7 | 0.8 | 0.9 | 0.8 | 1 | 0.7 | 0.6 | 0.7 | 0.8 | 0.4 | 0.7 | 0.8 |
| $a_5$ | 0.2 | 0.4 | 0.7 | 0.8 | 0.6 | 0.3 | 0.9 | 0.4 | 0.4 | 0.8 | 0.1 | 1 |
| $a_6$ | 0.1 | 0.3 | 0.3 | 0.4 | 0.5 | 0.6 | 0.5 | 0.7 | 0.6 | 0.7 | 1 | 0 |
| $a_7$ | 1 | 0.9 | 0.8 | 0.8 | 0.7 | 0.3 | 0.2 | 0.8 | 0.7 | 0.2 | 0.8 | 0 |
| $a_8$ | 0 | 0 | 0.4 | 0.7 | 0.5 | 0.7 | 0.8 | 0.9 | 0 | 0.8 | 0 | 0.8 |

**Fig. 5.** Primary incidences matrix.

A single channel is enough to represent the single flow from an incident element towards an affected element. Even when all the incident elements that flow to an affected element are taken into account, the network of flows is as simple as follows in Fig. 6.

In this case of direct incidences, the human mind can easily estimate all the arcs in the network. In the hypothesis of the incorporation of the self-induced incidences, the brain "is thankful for" the help of an algorithm.

4. Construction of the fuzzy matrix $\left[\underset{\sim}{A}\right]$ of the valuations $(x_i, x_h)$ / i, h = 1, 2, ..., 8, of self-induced incidences of set A with itself (Fig. 7).
5. Obtaining the flow of semi-accumulated incidences using the max-min convolution operator.

The semi-accumulated incidences come together in a fuzzy matrix $\left[\underset{\sim}{M}'\right]$, resulting from the max-min convolution:

$$\left[\underset{\sim}{A}\right] \circ \left[\underset{\sim}{M}\right] \tag{9}$$

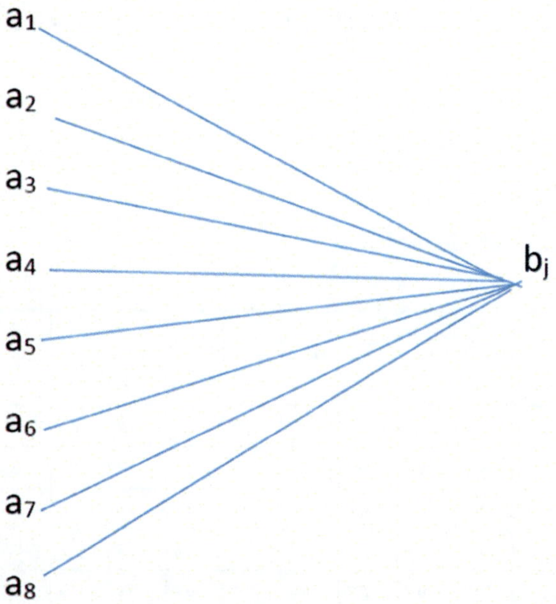


**Fig. 6.** Flow network from incident elements to an affected element.

$[\underset{\sim}{A}] =$

| $\underset{\sim}{A}$ | $a_1$ | $a_2$ | $a_3$ | $a_4$ | $a_5$ | $a_6$ | $a_7$ | $a_8$ |
|---|---|---|---|---|---|---|---|---|
| $a_1$ | 1 | 0.8 | 0.9 | 0.8 | 0.7 | 0.5 | 0.7 | 0.5 |
| $a_2$ | 0.6 | 1 | 0.5 | 0.5 | 0.4 | 0.4 | 0.5 | 0.7 |
| $a_3$ | 0.1 | 0.6 | 1 | 0.1 | 0.7 | 0.1 | 0 | 0.8 |
| $a_4$ | 0.3 | 0.7 | 0.8 | 1 | 0.8 | 0.4 | 0.7 | 0.8 |
| $a_5$ | 0.2 | 0.8 | 0.4 | 0.7 | 1 | 0.8 | 0.7 | 0.9 |
| $a_6$ | 0.3 | 0.5 | 0.2 | 0.1 | 0.1 | 1 | 0 | 0 |
| $a_7$ | 0.1 | 0.6 | 0 | 0.2 | 0 | 0.1 | 1 | 0.1 |
| $a_8$ | 0.8 | 0.7 | 0.6 | 0.4 | 0.9 | 0 | 0.2 | 1 |

**Fig. 7.** Self-induced incidences matrix of incident elements.

For this, for each pair we find $(x_i, y_j)$ i = 1, 2, …, 8: j = 1, 2, …, 12, the valuations:

$$(x_i, y_j)' = \bigvee_j \left((x_i, x_h) \bigwedge (x_h, y_j)\right) \qquad (10)$$

i, h = 1, 2, …8.
j = 1, 2, …, 12.
In our case, we obtain the Fig. 8.

$= [\underset{\sim}{A}] \circ [\underset{\sim}{M}]$

| $\underset{\sim}{A} \circ \underset{\sim}{M}$ | $b_1$ | $b_2$ | $b_3$ | $b_4$ | $b_5$ | $b_6$ | $b_7$ | $b_8$ | $b_9$ | $b_{10}$ | $b_{11}$ | $b_{12}$ |
|---|---|---|---|---|---|---|---|---|---|---|---|---|
| $a_1$ | 0.7 | 0.8 | 0.8 | 0.9 | 0.9 | 0.9 | 0.8 | 0.7 | 0.9 | 0.7 | 0.7 | 0.8 |
| $a_2$ | 0.6 | 0.7 | 0.5 | 0.7 | 0.6 | 0.8 | 0.7 | 0.7 | 0.5 | 0.7 | 0.5 | 0.8 |
| $a_3$ | 0.5 | 0.6 | 0.8 | 1 | 0.9 | 0.8 | 0.8 | 0.8 | 0.9 | 0.8 | 0.6 | 0.8 |
| $a_4$ | 0.7 | 0.8 | 0.9 | 0.8 | 1 | 0.8 | 0.8 | 0.8 | 0.8 | 0.8 | 0.7 | 0.8 |
| $a_5$ | 0.7 | 0.7 | 0.7 | 0.8 | 0.7 | 0.8 | 0.9 | 0.9 | 0.7 | 0.8 | 0.8 | 1 |
| $a_6$ | 0.5 | 0.5 | 0.3 | 0.4 | 0.5 | 0.6 | 0.5 | 0.7 | 0.6 | 0.7 | 1 | 0.5 |
| $a_7$ | 1 | 0.9 | 0.8 | 0.8 | 0.7 | 0.6 | 0.6 | 0.8 | 0.7 | 0.6 | 0.8 | 0.6 |
| $a_8$ | 0.6 | 0.8 | 0.7 | 0.8 | 0.6 | 0.8 | 0.9 | 0.9 | 0.6 | 0.8 | 0.6 | 0.9 |

**Fig. 8.** Semi-accumulated incidences matrix.

It is easy to see, and unsurprisingly, that the condition of the semi-convoluted matrix, $[\underset{\sim}{A}] \circ [\underset{\sim}{M}]$, contained in the fuzzy matrix of direct incidences, $[\underset{\sim}{M}]$, is satisfied.

6. Elaboration of the fuzzy matrix $[\underset{\sim}{B}]$ of self-induced incidences corresponding to the elements of the set of affected elements $b_j$/j = 1, 2, …, 12.

The information from the committee of experts allows us to bring together in matrix $[\underset{\sim}{B}]$ the valuations $(y_j, y_k)$/j, k = 1, 2, …, 12, as shown below in Fig. 9.

In relation to these valuations, the experts on the committee expressly informed us that in their opinion the criteria of incorporating in the valuations both the cases where the recycled or restored "thing" was used for its original purpose (entry into the circle) and when it was finally used as raw material or as a different product (exit from the circle) was prioritized.

With the construction of this matrix $[\underset{\sim}{B}]$ of self-induced incidences on the part of the primarily induced elements, $b_j$/j = 1, 2, …,12, we have all the numbered information needed to calculate the degree or level of total incidences.

$[\underset{\sim}{B}] =$

| $\underset{\sim}{B}$ | $b_1$ | $b_2$ | $b_3$ | $b_4$ | $b_5$ | $b_6$ | $b_7$ | $b_8$ | $b_9$ | $b_{10}$ | $b_{11}$ | $b_{12}$ |
|---|---|---|---|---|---|---|---|---|---|---|---|---|
| $b_1$ | 1 | 0.4 | 0.1 | 0.3 | 0.2 | 0.7 | 0.1 | 0.6 | 0.4 | 0 | 0.3 | 0 |
| $b_2$ | 0.4 | 1 | 0 | 0.6 | 0.1 | 0.3 | 0 | 0.2 | 0.1 | 0 | 0.7 | 0.2 |
| $b_3$ | 0 | 0 | 1 | 0.7 | 0.2 | 0.4 | 0 | 0.1 | 0.8 | 0 | 0.6 | 0 |
| $b_4$ | 0 | 0 | 0.7 | 1 | 0.6 | 0.8 | 0.2 | 0.1 | 0 | 0 | 0.8 | 0 |
| $b_5$ | 0 | 0 | 0.6 | 0.3 | 1 | 0.7 | 0.8 | 0 | 0 | 0 | 0 | 0.7 |
| $b_6$ | 0.6 | 0.7 | 0.8 | 0.8 | 0.7 | 1 | 0.2 | 0.7 | 0.8 | 0.2 | 0.8 | 0 |
| $b_7$ | 0.1 | 0.6 | 0.3 | 0.7 | 0.1 | 0 | 1 | 0.2 | 0.1 | 0 | 0.8 | 0 |
| $b_8$ | 0.4 | 0.6 | 0.3 | 0.2 | 0 | 0.7 | 0 | 1 | 0.2 | 0 | 0.8 | 0 |
| $b_9$ | 0 | 0 | 0.1 | 0.7 | 0.6 | 0.2 | 0.1 | 0.2 | 1 | 0.6 | 0.9 | 0 |
| $b_{10}$ | 0.1 | 0.2 | 0.1 | 0.4 | 0.1 | 0.1 | 0 | 0.3 | 0.8 | 1 | 0.8 | 0 |
| $b_{11}$ | 0.2 | 0.3 | 0.7 | 0.3 | 0.4 | 0.5 | 0.2 | 0.8 | 0.4 | 0.1 | 1 | 0 |
| $b_{12}$ | 0 | 0 | 0 | 0 | 0.6 | 0.7 | 0.8 | 0 | 0 | 0 | 0 | 1 |

**Fig. 9.** Self-induced incidences matrix of affected elements.

7. Obtaining the total accumulated flow of incidences:

Again, of course, we will use the max-min convolution operator to incorporate the flows of self-induced incidences of B, $\left[\underset{\sim}{B}\right]$ into those that already had them in the flow of semi-accumulated incidences $\left[\underset{\sim}{A}\right] \circ \left[\underset{\sim}{M}\right]$.

To do so, the following expression is used:

$$(x_i, y_i)^* = \bigvee_j \left((x_i, x_j)' \bigwedge (y_j, y_k)\right) \tag{11}$$

i = 1, 2, …, 8
j, k = 1, 2, …, 12.
We obtain the Fig. 10.

## 4.4 Degree or Level of the Flow Added by the Self-induced Incidences

A simple comparison of the direct incidences matrices $\left[\underset{\sim}{M}\right]$ and the $\left[\underset{\sim}{M}^*\right]$ of total incidences show us that in the different routes of flow, and in various of the channels, some differences are produced in the degree or level of the flow from an incident element to an affected element. To "number" this difference it is sufficient to use subtraction between the two matrices. In our example, the result was the Fig. 11.

The matrix $\left[\underset{\sim}{D}\right] = \left[\underset{\sim}{M}^*\right] (\text{-}) \left[\underset{\sim}{M}\right]$ found represents the addition of the flows of the self-induced incidences to the direct flows.

In this research, six major addition incidences were observed: $(x_2, y_4)$,, corresponding to the incidence of press articles and advertising on household items, the degree or level of direct incidence of which was 0.1, the total incidence going up to 0.8 as a result

$= [\underset{\sim}{A}] \circ [\underset{\sim}{M}] \circ [\underset{\sim}{B}]$

| $\underset{\sim}{A} \circ \underset{\sim}{M} \circ \underset{\sim}{B}$ | $b_1$ | $b_2$ | $b_3$ | $b_4$ | $b_5$ | $b_6$ | $b_7$ | $b_8$ | $b_9$ | $b_{10}$ | $b_{11}$ | $b_{12}$ |
|---|---|---|---|---|---|---|---|---|---|---|---|---|
| $a_1$ | 0.7 | 0.8 | 0.8 | 0.9 | 0.9 | 0.9 | 0.8 | 0.7 | 0.9 | 0.7 | 0.9 | 0.8 |
| $a_2$ | 0.6 | 0.7 | 0.8 | 0.8 | 0.7 | 0.8 | 0.8 | 0.7 | 0.8 | 0.7 | 0.8 | 0.8 |
| $a_3$ | 0.6 | 0.7 | 0.8 | 1 | 0.9 | 0.8 | 0.8 | 0.8 | 0.9 | 0.8 | 0.9 | 0.8 |
| $a_4$ | 0.7 | 0.8 | 0.9 | 0.8 | 1 | 0.8 | 0.8 | 0.8 | 0.8 | 0.8 | 0.8 | 0.8 |
| $a_5$ | 0.7 | 0.7 | 0.8 | 0.8 | 0.7 | 0.8 | 0.9 | 0.9 | 0.8 | 0.8 | 0.8 | 1 |
| $a_6$ | 0.6 | 0.6 | 0.7 | 0.6 | 0.6 | 0.7 | 0.5 | 0.8 | 0.7 | 0.7 | 1 | 0.5 |
| $a_7$ | 1 | 0.9 | 0.8 | 0.8 | 0.7 | 0.8 | 0.7 | 0.8 | 0.8 | 0.6 | 0.8 | 0.7 |
| $a_8$ | 0.6 | 0.8 | 0.8 | 0.8 | 0.7 | 0.8 | 0.9 | 0.9 | 0.8 | 0.8 | 0.8 | 0.9 |

**Fig. 10.** Total accumulated incidences matrix.

of incorporating flows of self-induced effects, providing the incorporation in the global flow with an incidence of 0.7; the incidence $(x_5, y_{11})$, recovery solidarity organisations on gift and decorative items, also with an incorporation in the global flow of 0.7; and the incidences $(x_8, y_2)$, $(x_8, y_9)$ and$(x_8, y_{11})$, illegal street selling on individual modes of transport, office equipment, and gift and decorative objects, respectively.

These last three accumulated incidences may seem strange. The human brain often forgets the effects that are presented to it in hidden form. To recover these forgotten items, we can use a reticular representation, where we can set a threshold of $\alpha \geq 0.8$ for the flows of incidences in the network.

$[\underset{\sim}{D}] =$

| $\underset{\sim}{D}$ | $b_1$ | $b_2$ | $b_3$ | $b_4$ | $b_5$ | $b_6$ | $b_7$ | $b_8$ | $b_9$ | $b_{10}$ | $b_{11}$ | $b_{12}$ |
|---|---|---|---|---|---|---|---|---|---|---|---|---|
| $a_1$ | 0.1 | 0 | 0.4 | 0.6 | 0.3 | 0 | 0 | 0.5 | 0.5 | 0 | 0.6 | 0.1 |
| $a_2$ | 0.1 | 0 | 0.5 | 0.7 | 0.3 | 0 | 0.1 | 0.3 | 0.3 | 0.1 | 0.6 | 0 |
| $a_3$ | 0.6 | 0.5 | 0 | 0 | 0 | 0 | 0.2 | 0.1 | 0 | 0.1 | 0.2 | 0.1 |
| $a_4$ | 0.2 | 0 | 0 | 0 | 0 | 0.1 | 0.2 | 0.1 | 0 | 0.4 | 0.1 | 0 |
| $a_5$ | 0.5 | 0.3 | 0.1 | 0 | 0.1 | 0.5 | 0 | 0.5 | 0.4 | 0 | 0.7 | 0 |
| $a_6$ | 0.5 | 0.6 | 0.4 | 0.2 | 0.1 | 0.1 | 0 | 0.1 | 0.1 | 0 | 0 | 0.5 |
| $a_7$ | 0 | 0 | 0 | 0 | 0 | 0.5 | 0.5 | 0 | 0.1 | 0.4 | 0 | 0.7 |
| $a_8$ | 0.6 | 0.8 | 0.4 | 0.1 | 0.2 | 0.1 | 0.1 | 0 | 0.8 | 0 | 0.8 | 0.1 |

**Fig. 11.** Flow added by the self-induced incidences matrix.

For greater visual clarity, we only consider the flows that start from an element of set A, for example, a8, illegal street selling, and pass through the network of channels, while the flow is equal to or greater than 0.8.

The corresponding network is shown below in Fig. 12.

A simple glance at this network is enough to make us realize that the total incidences of a8 "illegal street selling" on the elements $b_j/j = 1, 2, \ldots, 12$, effected elements, is very high, [0.8, 0.9], except for b1, cars, y b5, electronic household items, whose valuations, 0.6 and 0.7, respectively, are lower than the set threshold of $\alpha \geq 0{,}8$. The experts had only considered 4 of the 12 incident elements as incidences in this interval.

So, for:

- The direct incidence up to $b_2$, "personal modes of transport", a degree or level zero had been estimated, and through $a_1$, "government regulations and measures", an accumulated incidence of 0.8 is reached.
- The direct incidence up to $b_9$ "office equipment", the direct estimate is also zero and, through two paths $a_1$ "government regulations and measures" and $b_6$ "plastic products", an accumulated incidence of 0.8 is reached.

A reticular structure like this invites a wide range of considerations and interesting reflections.

## 4.5 A Future for the Circular Economy

Into the work we are just finishing we have tried to incorporate some theoretical and technical elements we have previously used in relation to other aspects of life in society into conceptual and methodological studies of the circular economy.

The reader will have observed that to do so we have delved deeply into this way of conceiving the economic science we have named humanism.

And, unlike those who understand economic humanism as an activity in favor of humankind playing the lead in their own story, we are convinced that humanism can only reach its true meaning if in handling the realities faced by man we use not only concepts but also techniques and operators capable of numerically expressing subjectivity.

This is the way we have generally approached our work, and the way in which we have approached this study.

We considered it was appropriate to present this brief and timely commentary because in many cases the human force of the object of study itself, in this case the circular economy, can divert attention and even hide the nature of its formulating process.

In the final analysis, we believe that the study of a human phenomenology can only happen with techniques and instruments capable of representing subjectivity.

There is, therefore, a certain logic in the fact that the choice of our central object of study of the circular economy has precisely been its circulatory scheme. But as we said previously, the conceptual, theoretical, and technical elements contributed can open up play and provide new perspectives in other areas of the circular economy.

It is strange that today, 5 June 2021, while I am writing down these ideas, the world is celebrating "Environment Day".

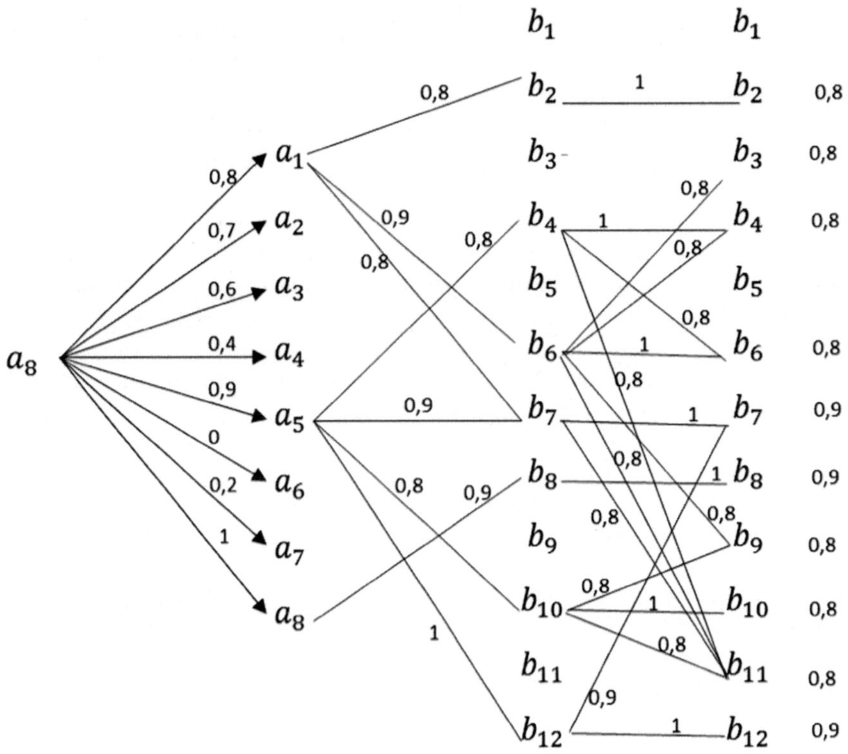


**Fig. 12.** Flow network from one incident elements to all the incided elements.

The visual, oral, and written media are comprehensively informing the world of the effects of climate change: the burning of fossil fuels, deforestation, contaminating spillages, and non-sustainable farm production and agriculture, with the consequent greenhouse gas emissions and the destruction of marine and land ecosystems, felt by populations in the form of floods, droughts, heatwaves, coldwaves, forest fires, and melting icecaps, all extreme meteorological phenomena.

Oppositely, possible actions to reduce emissions are also spoken of, among them the elimination of single-use plastics, greater consumption of electricity generated using renewable resources, reduced coal extraction, and the general minimization of carbon dioxide emissions.

All very good as elements of a plot of intertwining factors which, in one way or another, intervene to create the scene of a dramatic play: a future with no life on the planet.

But what can science in general do to help? Or even economic science with which we are concerned, or the circular economy which is our current interest?

The answer is a lot, at every level.

As a priority, classify and structure the problems that comprise this "all perceived". And, with a practical spirit, separate the various aspect to find scientific solutions one by one.

This is what we have started to do in the more operative part of this work, culminating in obtaining and describing the phases or stages, where appropriate, of an algorithm designed to determine, in the activities that make up the circle, the degree or level of incidence of one on the other.

And all of this with the objective of being able to take the best possible decisions and meet the desired objectives.

We believe we have achieved this by using elements of Artificial Intelligence to create a humanist algorithm. An algorithm whose functioning we have tried and tested.

But looking towards the future, we see other areas of economic activity related to the circular economy in which we can improve the results by means of decisions supported in the new discoveries of Artificial Intelligence, using humanist algorithms.

We will now take the privilege of citing some of those in which we have intervened to varying degrees.

Some studies that have had an impact are the ones on the "structure and functioning of the central circle and external incidences"; the de-termination of the degree or level of the "set of elements resulting from the circular economy" on the "set of configuring elements of climate change"; the degrees and levels of incidence of the "configuring elements of climate change on its specific consequences".

Possible quantification through measures and/or valuations is an essential source for establishing programs not only in the short term, but more importantly in the medium and long terms, so that by changing the direction of the destruction in which we are forced to take part, we are working to ensure that our planet continues to be a safe refuge for future generations of free citizens to live in in society.

Thank you very much.

## References

1. Zadeh, L.A.: Fuzzy Sets. Inf. Control **8**, 338–353 (1965)
2. Gil Aluja, J.: Papel de la memoria en la armonía entre territorios: el algorithm de Portugal. In: Complejidad Económica: una Península Ibérica más unida para una Europa más fuerte, RACEF, Barcelona (2019)
3. Gil Aluja, J.: Un ensayo para la solución al problema migratorio a través de la Inteligencia Artificial. In: Migraciones. RACEF, Barcelona (2020)
4. Gil Aluja, J.: Vejez y revolución digital. In: La vejez: conocimiento, vivencia y experiencia. RACEF, Barcelona (2021)
5. Gil Aluja, J.: Elements for a Theory of Decision in Uncertainty. Kluwer Academic Publishers, Dordrecht, Boston, London (1999)
6. Gil Aluja, J.: En el horizonte del poshumanismo. In: Desafíos de la nueva sociedad sobrecompleja: humanismo, transhumanismo, dataismo y otros ismos. RACEF, Barcelona (2019)
7. Kaufmann, A., Gil Aluja, J.: Grafos neuronales para la economía y gestión de empresas. Pirámide, Madrid (1995)
8. Gil Aluja, J.: Personal contribution for a new theory: the theory of self-induced incidences. In: Chivu, L., Franc, V.I., Georgescu, G., Andrei, J.V. (eds) Tangible Intention Assets in the context of European Integration and Globalisation Challenges ahead, pp. 70–71. Peter Lango, Berlín (2021)
9. Kaufmann, A., Gil Aluja, J.: Models per a la recerca d'efectes oblidats. Milladoiro, Vigo (1988)

# Finance and Economy

# Wavelet Entropy and Complexity Analysis of Cryptocurrencies Dynamics

Victoria Vampa[1], María T. Martín[2], Lucila Calderón[1,2], and Aurelio F. Bariviera[3](✉)

[1] Facultad de Ingeniería, Universidad Nacional de La Plata, La Plata, Argentina
[2] Facultad de Ciencias Exactas, Universidad Nacional de La Plata, La Plata, Argentina
[3] Department of Business, Universitat Rovira I Virgili, Av. Universitat 1, 43204 Reus, Spain
aurelio.fernandez@urv.cat

**Abstract.** Cryptocurrencies emerged almost one decade ago, as an alternative peer-to-peer payment method. Even though their currency characteristics have been challenged by several researchers, they constitute an important speculative financial asset. This paper examines the long memory properties in high frequency (5 min) time series of eight important cryptocurrencies. We perform a statistical analysis of two key financial characteristics of time series: return and volatility. We compute information theory quantifiers using a wavelet decomposition of the time series: wavelet entropy and wavelet statistical complexity of returns and volatility of each time series. We find two important features in the time series: (i) high frequency returns exhibit a trend toward a more efficient behavior, and (ii) high frequency volatility reflects a strong persistence in volatility. Both findings have important implications for portfolio managers, and investors in general. The presence of persistent volatility validates the use of GARCH-type models. Thus, understanding volatility could create opportunities for short-term day traders.

**Keywords:** Cryptocurrencies · Long memory · Statistical complexity · Wavelet entropy

## 1 Introduction

Eleven years ago, a white paper (anonymously posted online) proposed a new paradigm for validating financial transactions [1]. This new financial clearing setting, called 'blockchain', replaces the central trusted authority by a community validation based on consensus. Such validation uses a distributed ledger, which is replicated in servers within the network. The existence of several simultaneous encrypted copies of the transaction history produced robustness against data manipulation. Thus, it constitutes a safe alternative to validating transactions. The vehicle in such transactions is a new asset called cryptocurrency, due to its encrypted nature. Bitcoin price rocketed during 2017, reaching in December of that year almost $20,000 [2]. If we assume that an investor bought one bitcoin in June 2009 at around $0.0001, he would have earned approximately 847% compounded annually. Such success encouraged other software developers to create new cryptocurrencies. Until 2016 Bitcoin was the unchallenged dominant player, with more


M. d. P. Rodríguez García et al. (Eds.): XX SIGEF 2021, LNNS 384, pp. 25–35, 2022.
https://doi.org/10.1007/978-3-030-94485-8_2

than 90% of the market. As of July 2020, there are more than 5000 cryptocurrencies (coins and tokens), with a market capitalization of $272 billion and daily transactions of around $56 billion. Bitcoin represents 62% of the whole market [3].

This market is very dynamic with new coins (globally known as 'altcoins') entering the market every week, and some coins stopping trading because of lack of investors' attention.

Despite its name, cryptocurrencies are more an actively traded financial asset rather than a standard currency. Its daily use is really very limited. Coinmap [4] reports that there are only 22891 physical venues in the world accepting payments in cryptocurrencies, limiting their use as a medium of exchange. Volatility of crytocurrencies is ten times higher than any other financial assets such as fiat currencies, bonds, stocks or precious metals. Moreover, volatility exhibits long term memory [5]. Such situation challenges the use of cryptocurrencies use as store of value.

The aim of this paper is to revisit two important empirical features of cryptocurrencies time series: (i) the informational efficiency, and (ii) the persistence in volatility. The period under study is very relevant, because it covers the final period of the price bubble formed in the last quarter of 2017 and the aftermath of the bubble burst.

Our paper contributes to the existing literature in several ways. First, we study high frequency time series of the main cryptocurrencies, whereas most of the papers focused only on daily returns. Second, our technique (wavelet entropy) has been not used yet in this market. Third, we study not only high frequency returns, but also high frequency volatility in the period around the bubble burst of 2017–2018. This approach is relevant to provide an assessment of market risk.

The paper is organized as follows. Section 2 reviews the most relevant cryptocurrency literature. Section 3 presents the methodology. Section 4 describes data and discusses the main findings. Finally, Sect. 5 outlines the conclusion of our analysis.

## 2 Brief Literature Review

Scientific literature around cryptocurrencies covers different facets of the problem. From an economics point of view, an important discussion deals with the characterization of cryptocurrencies as "currencies". Yermack [6] argues that Bitcoin does not reach the currency status, as it performs poorly as a unit of account and as a store of value. Several papers [7, 8] characterize bitcoin mainly as a speculative asset rather than a currency. It is widely recognized [9–11] that bitcoin (and generally most cryptocurrencies) conform a unique and uncorrelated asset class, compared with traditional assets such as stocks, bonds, or commodities.

One important research line within financial economics is the study of the informational endowment of cryptocurrencies time series. The seminal paper by [12] defines informational efficiency as a situation when market prices fully reflect all available information. It is conventionally classified into three broad categories (weak, semi-strong, and strong), depending on the set of information considered as benchmark. This paper studies the weak form of informational efficiency, meaning that returns should follow a white noise.

Although the informational in efficiency of the cryptocurrency market is well-documented [5, 13–15], most studies use daily data. Even though, there are papers that use intraday data [10, 16–18], their focus is exclusively on bitcoin.

## 3 Wavelet Entropy and Wavelet Complexity

Financial markets, and in our case cryptocurrencies' markets, register each transaction, including attributes such as time, price, and quantity of the operation. A close scrutiny on these data could be useful to extract information on the statistical characteristics of the price generating process in a given market.

Information theory-based quantifiers could be a suitable alternative to more traditional econometric methods in time series analysis. Entropy quantifiers have a long tradition in economics. Three papers dated in the 1960s were among the first to use information entropy to examine the predictability and temporal dependence in stock time series [19–21].

The celebrated Shannon entropy [22] constitutes a straightforward way to measure the degree of (dis)order in a system. Let $P = \{p_i \in \mathbb{R}; \mathrm{p_i} \geq 0; \mathrm{i} = 1, \ldots, \mathrm{M}\}$, be a discrete probability distribution, with $\sum_{i=1}^{M} p_i = 1$, then the Shannon entropy reads:

$$S[P] = \sum_{i_1}^{M} p_i \log p_i \quad (1)$$

This quantifier equals zero if the patterns are fully deterministic and reaches its maximum value for a uniform distribution.

Wavelet entropy is based on Shannon entropy and is computed by defining a probability distribution that arises from the multiresolution analysis on $L^2(\mathbb{R})$ [23]. Given a multiresolution analysis $\{V_j\}_{j \in Z}$, and being $W\_j$ the orthogonal complement of $V_{j+1}$ in $V_j$ with:

$$W_j \oplus V_j = V_{j+1} \quad (2)$$

it is obtained that:

$$L^2(\mathbb{R}) = \cdots \oplus W_{-1} \oplus W_0 \oplus W_1 \oplus \quad (3)$$

Then, the original signal $\mathrm{s} \in \mathrm{L}^2(\mathbb{R})$, could be written as:

$$\mathrm{s(t)} = \sum_{\mathrm{j} \in \mathbb{Z}} \sum_{\mathrm{k} \in \mathbb{Z}} \mathrm{d_j(k)} \psi_{\mathrm{j,k}}(\mathrm{t}) \quad (4)$$

where $\{\psi_{\mathrm{j,k}}\}$ for each given $j$, is a base on the space $W_j$.

The probability distribution is obtained from the concept of "wavelet energy":

$$\mathrm{E_j} := \sum_{\mathrm{k} \in \mathbb{Z}} |\mathrm{d_j(k)}|^2 \quad (5)$$

Consequently, the total energy of a signal $s(t)$ can be obtained by adding up the energy of all levels $j$:

$$\mathrm{E_{total}} := \sum_{\mathrm{j} \in \mathbb{Z}} \sum_{\mathrm{k} \in \mathbb{Z}} |\mathrm{d_j(k)}|^2 \quad (6)$$

Using Eqs. (5) and (6), we define the energy probability distribution as:

$$p_j := \frac{E_j}{E_{total}} \quad (7)$$

Finally, the wavelet entropy $S_W[P]$ is defined using these $p_j$ in Eq. (1).

Another information quantifier is the Statistical Complexity Measure (SCM), which is a global information quantifier [24], and will be denoted by C[P].

For a discrete probability distribution function $P = \{p_i \in \mathbb{R}; p_i \geq 0; i = 1, \ldots, M\}$, associated with a time series, this functional $\mathbf{C}[P]$ is given by

$$C[P] = H[P] \cdot Q[P] \quad (8)$$

where $H[P]$ is the normalized Shannon entropy and Q[P] is the disequilibrium-distance.

$H[P]$ measures the amount of disorder and is defined as,

$$H[P] = \frac{S[P]}{S_{max}} \quad (9)$$

where S[P] is the Shannon entropy of Eq. (1) and $S_{max} = S[P_e]$ corresponds to the uniform distribution $P_e = \{p_i = \frac{1}{M}; i = 1, \ldots, M\}$, and so has the value $S_{max} = \log(M)$.

Q[P] is called the disequilibrium-distance. It is defined as:

$$Q[P] = Q_0 \cdot D[P, P_e] \quad (10)$$

where $Q_0$ is a normalization constant ($0 \leq Q \leq 1$) with its value equal to the inverse of the maximum possible value of the distance $D[P, P_e]$. This maximum distance is obtained when one of the components of $P$, say $p_m$ is equal to one, while the remaining components are equal to zero. The disequilibrium-distance **Q** would be different from zero if there exist more likely states among the accessible ones.

Several distance-forms open several possibilities for the SCM [24]. In this work we consider the disequilibrium form for the complexity measure proposed by Lopez Ruiz, Mancini and Calbet (LMC-complexity measure) in which the Euclidean norm is considered:

$$D_E\left[P^{(1)}, P^{(2)}\right] = |P^{(1)} - P^{(2)}|_E^2 = \sum_{j=1}^{M} \left(p_j^{(1)} - p_j^{(2)}\right)^2 \quad (11)$$

with $P^{(l)} = \left\{p_i^{(l)} \in \mathbb{R}; p_i \geq 0; i = 1, \ldots, M\right\}$, $l = 1, 2$ are two discrete probability distributions.

Giving a measure of the complexity of a time series, $\mathbf{C}[P]$ quantifies the existence of correlational structures. In perfect order or total randomness of the signal, the value of $\mathbf{C}[P]$ is identically zero meaning the signal possesses no structure. A large range of possible stages may be realized between these two extreme instances and quantified by a nonzero $\mathbf{C}[P]$.

The statistical complexity does not only quantify randomness (or disorder) but also the degree of correlation between structures. It is a non-trivial function of entropy, and it is important to note that, for a given value of $H[P]]\$$ there is a range of possible values for $\mathbf{C}[P]$ between a minimum and a maximum value [24].

Once these quantifiers $H[P]$ and $\mathbf{C}[P]$ are calculated, the results can be displayed in the $H\mathbf{C}$ plane. In this way, specific features associated with the dynamics of the series can be characterized.

In the present work, a new approach for the characterization of returns and volatility time series is presented. The information theory quantifiers $\mathrm{H_W[P]}$ and $\mathrm{C_{LMC}[P]}$ were used. The first one,$H_W[P]$, uses wavelet entropy $\mathrm{S_W[P]}$ in Eq. (9), while the complexity quantifier $\mathrm{C_{LMC}[P]}$ is obtained from Eq. (8) using the Euclidean distance $\mathrm{D_E}$ in Eq. (11) to measure the disequilibrium-distance, Eq. (10).

# 4 Data and Results

The empirical analysis was performed using high frequency (5 min) price data from eight of the main cryptocurrencies. In this way, signals $s(t)$ have returns and volatility values corresponding to a uniform time grid.

Data coverage spans from 03/12/2017 until 12/04/2018, comprising 29175 observations. Data were retrieved from Eikon Thomsom Reuters. The selection of the data span was justified because it included a period of boom and bust in the cryptocurrency market. During December 2017 there was an unprecedented increase in cryptocurrencies, followed by a crash during the first months of 2018. In this sense. Contrary to previous studies, we would like to focus our study during a time of market stress.

Following [5], we adopt a dynamic approach and compute the information quantifiers (wavelet entropy and LMC statistical complexity) using rolling windows. This procedure works as follows: we consider the first 128 observations (roughly half a day) and compute the quantifiers. Then we select the following 128 observations and compute their respective quantifiers. We repeat the algorithm until the end of the time series. We obtained 227 rolling windows.

Our study detects several interesting features in the cryptocurrency ecosystem, regarding both returns and volatility.

## 4.1 Cryptocurrencies' Returns

Our initial analysis explores the long memory of the time series, using the wavelet entropy-complexity plane. This representation allows to characterize different stochastic and chaotic dynamics. For the purpose of identification, we draw the confidence interval of the location of a Brownian motion with Hurst exponent equal to 0.5[1].

[1] This confidence interval was obtained by simulating 1000 artificial time series using the function wfbm from Matlab and computing the average and standard deviation of wavelet entropy and statistical complexity.

Figure 1 shows how the time series of all the cryptocurrencies occupy a broad extension of this plane. This situation means that the time series goes through different stages in stochastic dynamics. In fact, only a small proportion of the sliding windows lies in the area corresponding to Hurst = 0.5. We can interpret that most of the time, the time series exhibit an informational inefficient behavior. One drawback of this representation is that it does not allow to identify how the position of the quantifiers evolve in time.

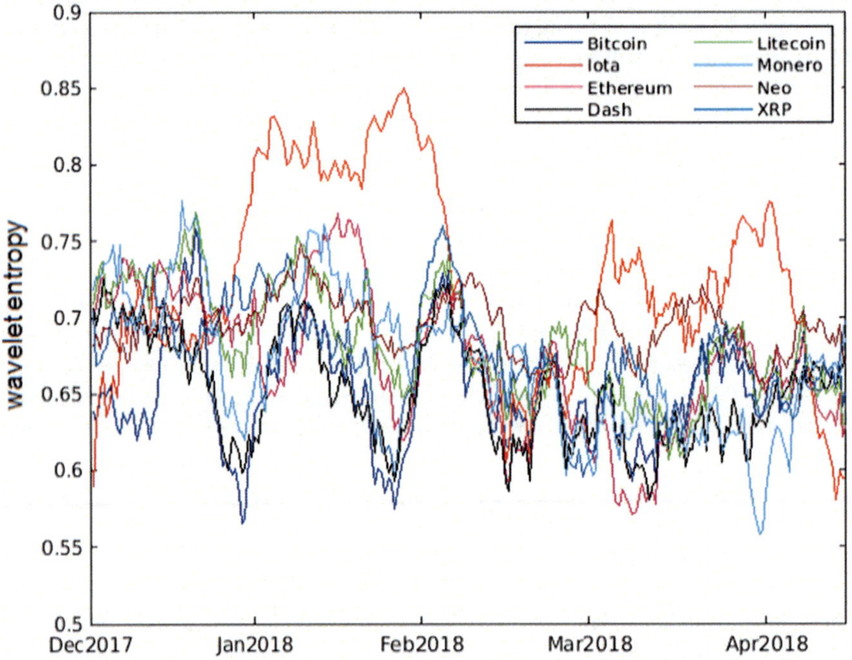


In order to overcome this situation, we present also Fig. 2, which describe the evolution of the wavelet entropy along the different sliding windows. This figure reveals that the informational efficiency, proxied by the wavelet entropy, varies significantly. There are periods when the time series exhibit a fairly random behavior, whereas in other periods the time series become informational inefficient. However, there is a clear trend towards a more efficient behavior.

Finally, this figure uncovers a remarkable variety in the cryptocurrency market. The movements in wavelet entropy are not completely synchronized for the different currencies under analysis. However, it emerges a trend to a closer comovement in this quantifier by the end of the examining period. One striking feature, that was not reported in previous literature, is that IOTA exhibits a dynamic that is apart from the other coins. Altogether these results have two important implications from a portfolio perspective. On the one hand, if cryptocurrencies display a convergence in the stochastic dynamics, means that the market is more closely integrated. On the other hand, the existence of some cryptocurrencies (in our study, IOTA), with mismatched dynamics could be used to exploit diversification benefits in portfolios.

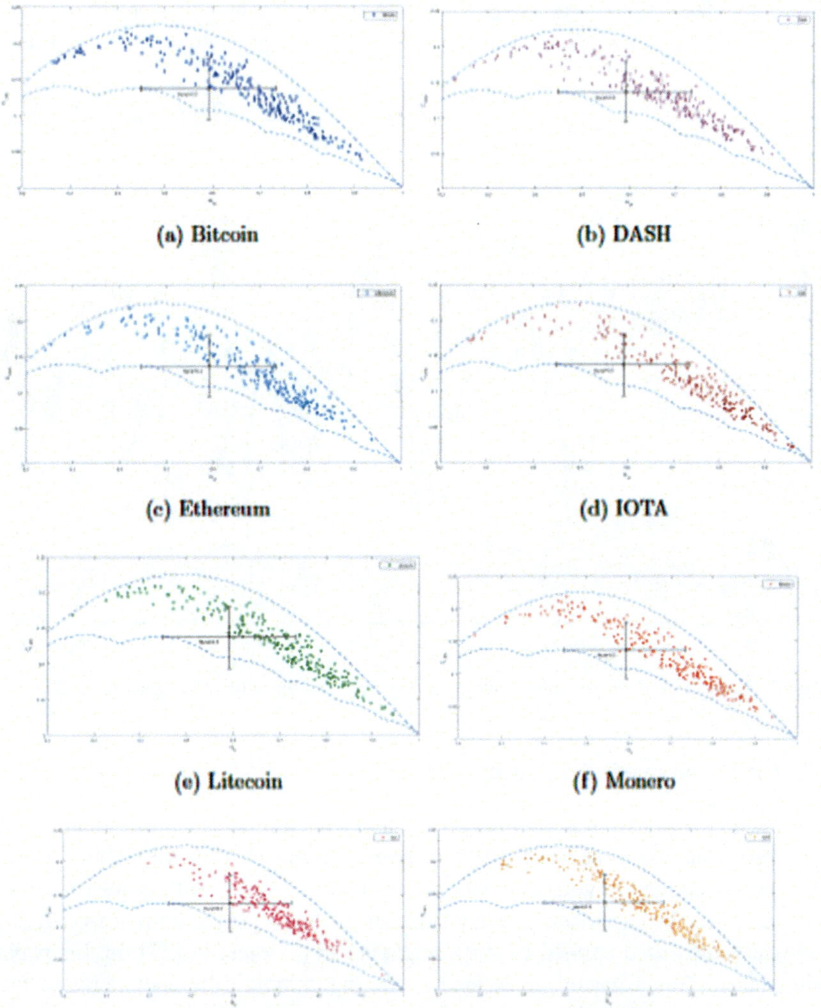

(a) Bitcoin (b) DASH

(c) Ethereum (d) IOTA

(e) Litecoin (f) Monero

(g) Neo (h) Ripple

**Fig. 1.** Wavelet entropy-complexity plane of cryptocurrencies' returns

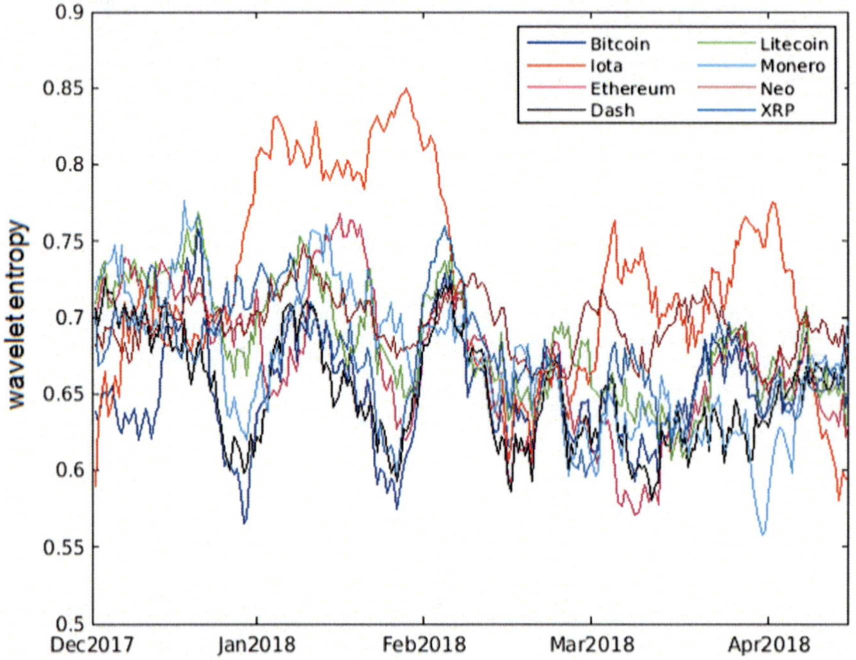


**Fig. 2.** Evolution of the wavelet entropy across time for the selected cryptocurrencies' returns

## 4.2 Cryptocurrencies' Volatility

It has been previously reported by [11] that cryptocurrencies exhibit greater volatility than traditional assets such as stocks or bonds. In this section we attempt to measure the memory content in volatility time series. There several volatility measures, such as those proposed by Parkinson [27] and Garman and Klass [28]. A comparison of different alternative measured applied to stock markets can be found in [29]. In this paper, and following [5, 30], volatility is defined as the logarithmic difference between the high and low values recorded within the 5 min interval of each observation. Considering that cryptocurrencies are, mainly, a speculative asset, it attracts short-term, who look to exploit price changes that occur minute to minute. Therefore, studying the volatility profile is crucial for developing a profitable trading strategy.

In this line, another important finding is that all the time series under analysis exhibit strong persistent volatility, reflecting Hurst exponents between 0.8 and 0.9. Such situation means that periods of high (low) volatility are more likely to be followed by periods of high (low) volatility (See Fig. 3). Then, daily traders could formulate algorithmic trading strategies, in order to take advantage of uninformed traders.

From a theoretical point of view, this characteristic should be taken into account when modeling volatility. Such persistence favors the consideration of fractional integrated and hyperbolic GARCH models, such as those proposed by [25] or [26], to allow the presence of long-range dependence in volatility.

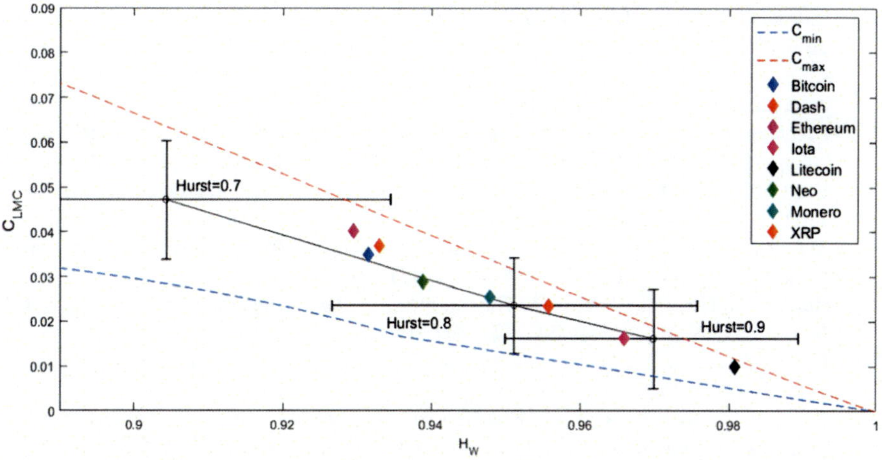


**Fig. 3.** Wavelet entropy-complexity Plane for cryptocurrencies' volatilities

## 5 Conclusions

This paper studies the persistence in the time series of eight cryptocurrencies return and volatility time series. Unlike previous literature, we study the stochastic behavior of high frequency time series. This analysis is justified in the existence of day traders, who by means of automatic intelligent algorithms, try to discover profitable trading patterns to be exploited in the very short run. Another difference with the previous literature is that we focus our analysis on a particular time frame from December 2017 until early April 2018. This period comprises a great boom and a subsequent bust in cryptocurrency prices. In this sense, the paper sheds light on the behavior of the long memory under stressing market conditions.

In this paper we compare the information endowment of the time series using a rolling window approach. We detect that, in the period under examination, the time series exhibit both efficient and inefficient behavior. However, there is a remarkable trend toward a more efficient behavior. We also argue that the market is heterogeneous in terms of stochastic dynamics, albeit it is moving towards a more synchronizing behavior.

A significant result is regarding volatility. We find that volatility is highly persistent for all the coins in our sample, and during the whole period. As a consequence, speculative traders could find this market particularly attractive, considering that they could test their trading algorithms seven days a week and at any time.

Finally, we detect that IOTA seems to have a distinct dynamic. These results could be important for portfolio managers, interested in forming port- folios with different coins, because they can control the overall risk by diversifying in different cryptocurrencies.

## References

1. Nakamoto, S.: Bitcoin: A peer-to-peer electronic cash system. https://bitcoin.org/bitcoin.pdf/. Accessed 27 Dec 2019 (2009)

2. Morris, D.Z.: Bitcoin hits a new record high, but stops short of $20,000. https://fortune.com, 2017/12/17/ bitcoin-record-high-short-of-20000/. Accessed 16 Jul 2020 (2017)
3. Coinmarket: Crypto-Currency Market Capitalizations. https://coinmarketcap.com/currencies/. Accessed 16 Jul 2020 (2019)
4. Coinmap: All the cryptocurrency merchants and ATMs of the world in one map. https://coinmap.org/view/#/world/34.16181816/-13.18359375/2. Accessed 18 Jul 2020 (2020)
5. Bariviera, A.F.: The inefficiency of Bitcoin revisited: a dynamic approach. Econ. Lett. **161**, 1–4 (2017)
6. Yermack, D.: Is bitcoin a real currency an economic appraisal, NBER Working Paper Series. https://www.nber.orgpapers/w19747.pdf (2013)
7. Baur, D.G., Hong, K., Lee, A.D.: Bitcoin medium of exchange or speculative assets. J. Int. Finan. Markets. Inst. Money **54**, 177–189 (2018)
8. Corbet, S., Lucey, B., Peat, M., Vigne, S.: Bitcoin futures—what use are they. Econ. Lett. **172**, 23–27 (2018)
9. Bouri, E., Molnár, P., Azzi, G., Roubaud, D., Hagfors, L.I.: On the hedge and safe haven properties of bitcoin: is it really more than a diversifier? Financ. Res. Lett. **20**, 192–198 (2017)
10. Corbet, S., Meegan, A., Larkin, C., Lucey, B., Yarovaya, L.: Exploring the dynamic relationships between cryptocurrencies and other financial assets. Econ. Lett. **165**, 28–34 (2018)
11. Aslanidis, N., Bariviera, A.F., Martinez-Ibañez, O.: An analysis of cryptocurrencies conditional cross correlations. Financ. Res. Lett. **31**, 130–137 (2019)
12. Fama, E.F.: Efficient capital markets: a review of theory and empirical work. J. Financ. **25**(2), 383–417 (1970)
13. Urquhart, A.: The inefficiency of Bitcoin. Econ. Lett. **148**, 80–82 (2016)
14. Nadarajah, S., Chu, J.: On the inefficiency of Bitcoin. Econ. Lett. **150**, 6–9 (2017)
15. Bariviera, A.F., Basgall, M.J., Hasperué, W., Naiouf, M.: Some stylized facts of the Bitcoin market. Physica A **484**, 82–90 (2017)
16. Takaishi, T., Adachi, T.: Taylor effect in bitcoin time series. Econ. Lett. **172**, 5–7 (2018)
17. Li, X., Li, S., Xu, C.: Price clustering in bitcoin market—an extension. Financ. Res. Lett. **32**, 101072 (2020)
18. Sensoy, A.: The inefficiency of bitcoin revisited: a high-frequency analysis with alternative currencies. Financ. Res. Lett. **28**, 68–73 (2019)
19. Fama, E.F.: Tomorrow on the New York stock exchange. J. Bus. **38**(3), 285–299 (1965)
20. Theil, H., Leenders, C.T.: Tomorrow on the Amsterdam stock exchange. J. Bus. **38**(3), 277–284 (1965)
21. Dryden, M.M.: Short-term forecasting of share prices: an information theory approach. Scot. J. Polit. Econ. **15**(1), 227–249 (1968)
22. Shannon, C.E., Weaver, W.: The Mathematical Theory of Communication. University of Illinois Press, Champaign, IL (1949)
23. Chui, C.: An Introduction on Wavelet Analysis. Academic Press, New York (1992)
24. Kowalski, A.M., Martín, M.T., Plastino, A., Rosso, O.A., Casas, M.: Distances in probability space and the statistical complexity setup. Entropy **13**(6), 1055–1075 (2011)
25. Baillie, R., Bollerslev, T., Mikkelsen, H.: Fractionally integrated generalized autoregressive conditional heteroskedasticity. J. Econometr. **74**(1), 3–30 (1996)
26. Davidson, J.: Moment and memory properties of linear conditional heteroscedasticity models, and a new model. J. Bus. Econ. Stat. **22**(1), 16–29 (2004)
27. Parkinson, M.: The extreme value method for estimating the variance of the rate of return. J. Bus. **53**(1), 61–65 (1980)
28. Garman, M.B., Klass, M.J.: On the estimation of security price volatilities from historical data. J. Bus. **53**(1), 67–78 (1980)

29. Floros, C.: Modelling volatility using high, low, open and closing prices: evidence from four S&P indices. Int. Res. J. Financ. Econ. **28**, 198–206 (2009)
30. Bariviera, A.F., Zunino, L., Rosso, O.A.: An analysis of high-frequency cryptocurrencies prices dynamics using permutation-information-theory quantifiers. Chaos **28**(7), 075511 (2018)

Check for updates

# The Use of Fuzzy Decoupled Net Present Value in Pepper Production

José M. Brotons-Martínez[1] (✉), Amparo Galvez[2], Ruben Chavez-Rivera[3], and Josefa Lopez-Marín[2]

[1] Department of Economic and Financial Studies, Miguel Hernández University, 03202 Elche, Alicante, Spain
jm.brotons@umh.es
[2] Department of Crop Production and Agri-Technology, IMIDA, 30150 Murcia, Spain
[3] Facultad de Químico Farmacobiología, Universidad Michoacana de San Nicolás de Hidalgo, Morelia, Michoacán, México

**Abstract.** Pepper crops are, after tomatoes and cucumbers, the third-largest cultivated area of greenhouse crops. Its cultivation presents not only economic risks, especially inherent to prices fluctuations and costs rising, but also agronomy risks such as the presence of nematodes, aphids or water salinity. Traditionally discounted cash flows have been used to analyze the viability of these types of crops. Recently, decoupled net value was used to improve this valuation. In this paper, we propose the use of Fuzzy Decoupled Net Present Value. Once we have obtained the historical or bibliographic information, it has been presented to experts who have to report the risk in the area. This price risk is not the same anywhere, and the water salinity risk depends on the place where the farmer is going to grow peppers. Risk evaluation has been carried out through the analysis of pepper plantations for 2016 and 2017 using fuzzy mathematics. The use of decoupled net present value eliminates the risk premium interest rate and has permitted an increase in the accuracy and quantification of risks, isolating the main risks such as price drops, temporally and definitive (EUR 7680 $ha^{-1}$ $year^{-1}$) or salinity risks (EUR 4640 $ha^{-1}$ $year^{-1}$). The use of decreasing discount functions has permitted a more realistic investment estimation.

**Keywords:** Pepper · Risk · Fuzzy decoupled net present value · Price fall

## 1 Introduction

Pepper crops are highly important globally and they occupy the third largest cultivated area of greenhouse crops, after tomatoes and cucumbers, but they are second in terms of economic importance. Spain is considered the fifth largest pepper-producer in the world (1,275,457 tons in 2018, and the second exporter behind Mexico with 775,771 tons [1]). Due to overlapping production calendars, Turkey is Spain's main competitor [2]. Currently, pepper production and marketing are conditioned by economic, environmental, and quality aspects.


M. d. P. Rodríguez García et al. (Eds.): XX SIGEF 2021, LNNS 384, pp. 36–46, 2022.
https://doi.org/10.1007/978-3-030-94485-8_3

Although pesticides can also be used to achieve better production, their use is becoming increasingly limited because they cause environmental pollution as well as harm to human health and fauna [3], as seen in bees. Additionally, the fact that pests are becoming more and more resistant to the use of pesticides, causes a risk to crops. The fundamental risks to pepper crops from pests and diseases are biotic problems such as nematodes [3–6] or aphids [7, 8]. There are also abiotic problems caused by the shortage and salinity of water [9–11] due to climatic changes, among others.

Last but not least important is the risk arising from adverse movements in the selling prices of peppers. Growers establish their crops with the expectation of reaching prices that are similar to those from previous campaigns. However, these prices can experience significant fluctuations that endanger the profitability of current as well as future production campaigns [12, 13], which can even lead to growers abandoning the crop.

The use of net present value (NPV) stands out among the traditional methods for evaluating investment projects [14]. This method is based on the weighted average cost of capital (WACC), requiring the same return on investments as the cost of financing via equity or debt funds. The introduction of risk in investment projects is usually done by increasing the discount rate. However, this procedure generates different problems, such as penalizing long-term projects or those that present high values in net cash flows over time [15].

To solve these problems, the decoupled net present value introduced by Espinoza [16] first identifies the risks, then through probabilistic analysis combined with the valuation of options, it introduces the values of these risks as a cost of the project. However, there are several risks whose estimation is very hard according to historical data. That is why the use of fuzzy logic is especially interesting. The application of this branch of mathematics has been extended to several fields such as economy of agronomy. Fuzzy logic entails a proper treatment of subjectivity and uncertainty, and it also provides a series of instruments for analysts to be able to transmit the information [17]. Fuzzy methodology has been used to explain agronomic phenomena such as Brotons et al. [18] to analysis the loss of the green colour in lemon rind during ripening.

For all the above reasons, NPV will be used in combination with a correct analysis of the inherent risks of the operation using DNPV and fuzzy mathematics. In particular, an analysis will be made of the temporal and permanent fall in prices throughout the useful life of the installation. Other risks will also be analyzed, such as losses due to pests because of not being treated in time, or due to a poor response to phytosanitary treatments. In addition, risk due to a worsening in the quality of irrigation is introduced. All of the above will enable us to determine a present investment value that is very close to the real value.

## 2 Materials and Methods

The operator present value converts a cash flow into its present value and indicates that the one using this operator is indifferent to the general cash flow at time t and the present cash flow in zero $P\left[\tilde{c}_i, t_i\right] = \sum_i \tilde{c}_i d(t_i)$ where $d(t_i)$ is the discounting function. In this work, exponential functions are used, $d(t_i) = e^{-r_i t}$ being $r_i$ the instantaneous interest rate, and $c_i$ is the cash flow, and is obtained as the difference between income and foreseen payments. In most cases, future net cash flows ($\tilde{c}_i$) can be considered as uncertain. In any case, they can be defined as the difference between cash inflows $\tilde{I}_i$ and cash outflows $\tilde{H}_i$, which are also considered uncertain $\tilde{c}_i = \tilde{I}_i - \tilde{H}_i,\ i = 1, \ldots, n$.

In an environment of uncertainty, decoupled net present value will be used. In general, the problem with risk investments is due to the fact that the real value of future cash flows tends to be worse than the foreseen value. For this reason, future income and expense flow is reduced with the value of the corresponding synthetic insurance, which protects the investor from future risks of loss in value of future cash flows. In this sense, the value of the synthetic insurance of inflows is $\tilde{S}_{I_i}$, and that of outflows is $\tilde{S}_{H_i}$,

$$P[c_i, t_i] = \sum_i \left[\tilde{I}_i - \tilde{H}_i - \left(\tilde{S}_{I_i} - \tilde{S}_{H_i}\right)\right] \cdot e^{-r_i t} \quad (1)$$

In this way, by combining the discount function and decoupled net present value, it is possible to obtain a project's value in several years' time taking into account the real risk of future cash flows. Considering that the risk to analyse in this paper are very difficult to be evaluated only with historical information, the opinion of the experts will be considered. The use of fuzzy mathematics will be very suitable to deal with the subjectivity inherent in the experts' opinion. As a result, the present model will be called Fuzzy Decoupled Net Present Value (FDNPV). The risks to be evaluated are:

a) Temporary price fall. The risk of a temporary price fall has been calculated by considering that if the price does not fall below the profitability threshold, this decline can be considered circumstantial, and it is therefore foreseeable that in the following years, prices will recover. It is the risk that prices will not reach the expected values during the campaign. For this purpose, several experts from the area of cultivation will be consulted. The future price will be valuated according to the linguistic labels {very low, a few low, lightly low, lightly high, a few high, very high} corresponding to these prices ranges {0–0.15; 0–0.3; 0.15–0.45; 0.3–0.6; 0.45–0.75; 0.6–0.9}. Experts will be asked to assess the possibility the prices fall in each range. The final assessment will be done using the assessment of the linguistic labels of Table 1.

**Table 1.** Values assigned to the linguistic labels

| Linguistic label | $v_j$ |
|---|---|
| (1): Totally disagree | 0.00 |
| (2): Strongly disagree | 0.20 |
| (3): Disagree | 0.40 |
| (4): Agree | 0.60 |
| (5): Strongly agree | 0.80 |
| (6): Totally agree | 1.00 |

The membership for each range of prices has been obtained according to the following expression

$$\mu_i = \frac{1}{m}\sum_{j=1}^{6} \nu_i a_{ij} \tag{2}$$

Being i the price range assessed and m the number of experts who have answered each range of prices and $n_{ij}$ the number of experts that assess range i with the linguistic label j.

The probability that the price falls between the threshold (price cost) and the average market price can be obtained as the quotation of the sum of the membership functions of the linguistic labels between both prices and the total sum of the linguistic labels for the whole range. For those that include partially the threshold (or the market average), the possibility will be obtained proportionally. The average market price has been obtained from the information available at CARM [19].

The average price has been obtained by multiplying each linguistic label by its membership function. The loss will be obtained as the difference between the expected value (for the prices from the threshold to the average market price) and the threshold. Lastly, the risk (by hectare) is obtained by multiplying the loss by its probability and by the hectare production.

(b) Risk due to permanent fall in prices. The risk of a permanent fall in prices has been obtained in a similar way, but considering the possibility the price falls below the profit-ability threshold price.

(c) Risk due to existence of nematodes and aphids. Bibliography has been consulted to obtain the risks due to the most common and harmful pests in peppers [3–8], comparing the production of plants affected and unaffected by this circumstance. This information has been asked to the experts, and linguistic labels were used in a similar way to the previous case to estimate the probability of occurrence in the crop fields.

(d) Risk due to existence of bad quality irrigation water. For the risk of having to use water with a greater concentration of salt, which is harmful to plants, four scenarios (top quality, 2.5 $dSm^{-1}$, 3.5 $dSm^{-1}$, 6.0 $dSm^{-1}$) were considered according to the bibliography consulted [20, 21]. In these scenarios, production was compared to control production, and they were assigned probabilities of occurrence of episodes when waters have worse conditions. In a similar way to the previous cases, the membership function of each case has been obtained. The probability of each case has been obtained by dividing its membership function by the membership function sum of all the cases.

(e) Risk due to an increase in costs. Surveys of growers have been carried out, observing what increases in costs could occur and the occurrence probability, obtaining the risk of loss as a product of both, using linguistic labels in a similar way.

A prior NPV has been obtained without considering the risks previously calculated. This NPV has been obtained for each of the 24 years that the investment lasts and serves as a basis for obtaining the risk of a permanent fall in prices. Once this is obtained, all the risks obtained as an additional cost have been considered and the NPV of the project as well as yearly net yield have been obtained.

Interest rate has been considered as that of 10-year government bonds, which was 1.62% when this study was carried out [22].

# 3 Results

Mid-May is the start of the harvest of peppers cultivated in the Murcia region under thermal greenhouse covering. This means that two or three weeks in advance of the period they would be harvested without thermal covering. The mean price for this week was 1.04 € $kg^{-1}$ with a standard error of 0.10 € $kg^{-1}$. The price trend is clearly decreasing until it drops to prices around 0.45 € $kg^{-1}$, where it stabilizes until almost the end of the campaign.

## 3.1 Risk of Temporary Fall in Prices

During many campaigns, prices may not reach values considered as normal. This risk could be caused by circumstantial issues, such as occasional imbalances between supply and demand, which means that profit perspectives are not met for a specific year although the causes do not continue over time. This fall tends to be such that in spite of not obtaining the expected profit, the crop continues to be profitable for the grower.

Mean production under a micro tunnel greenhouse reaches 120,707 kg $ha^{-1}$, and a price of 0.378 € $kg^{-1}$ would cover costs. This means that the grower has to sell their product at higher prices to be able to cover at least the variable costs and part of the fixed costs or costs of setting it up. Conversely, the grower can be expected to abandon production.

The price the farmer will obtain during the year he will decide to grow the pepper is uncertain. Historical information is available, and in the best case, it is available weekly. However, no information is available for future prices. Although the estimation made from past information present good properties, it can be incomplete. That is why a group of five experts (two producers, one production technical, one researcher and one agricultural engineer) have been asked about his opinion about future prices. Six groups of prices have been considered (see Table 2) from 0.00 to 0.90 € $kg^{-1}$. Each expert has indicated his degree of agreement with the fact that price belongs to that range according to the value of Table 1. That is, for instance, for the range 0.00–0.15 they have to indicate if they are totally agree, strongly disagree, etc. The same for the range 0.00–0.30 € $kg^{-1}$, etc. Assigning the values of Table 1 to each linguistic label and adding the opinion of the experts, it is possible to obtain the membership function. Normalizing the membership function, it is possible to obtain the probability of each range.

Table 2 shows the number of experts who have assessed each range with each linguistic label. The last column is the probability that the price belongs to each range and has been obtained by dividing each membership function by the total sum of memberships.

The average market price for the last 10 years has been 0.5917 € $kg^{-1}$. Only lower prices will represent a loss for the farmer. In that case, if the price is higher than the threshold (0.3780 € $kg^{-1}$) the fall will be considered temporary, but if the price is lower than the threshold the fall will be considered permanent because the farmer may abandon the production. Now, we analyze the temporary fall. Table 3 shows the defuzzified value (central point) of each linguistic label and its membership function. Only those belonging to the range 0.3780–0.0591 € $kg^{-1}$ are considered. The expected value is 0.4070 € $kg^{-1}$. This is the price expected if there is a temporary fall. In that case, the expected loss is 0.185 € $kg^{-1}$ (the difference between the average market price 0.5917 € $kg^{-1}$ and

**Table 2.** Expert valuation about future prices

| Range | (1) | (2) | (3) | (4) | (5) | (6) | μ | Prob |
|---|---|---|---|---|---|---|---|---|
| 0.00–0.15 | 4 | 1 | 0 | 0 | 0 | 0 | 0.04 | 0.02 |
| 0.00–0.30 | 2 | 2 | 1 | 0 | 0 | 0 | 0.16 | 0.07 |
| 0.15–0.45 | 0 | 2 | 1 | 2 | 0 | 0 | 0.40 | 0.16 |
| 0.30–0.60 | 0 | 1 | 2 | 2 | 0 | 0 | 0.44 | 0.18 |
| 0.45–0.75 | 0 | 0 | 1 | 0 | 3 | 1 | 0.76 | 0.31 |
| 0.60–0.90 | 0 | 0 | 1 | 3 | 0 | 1 | 0.64 | 0.26 |

0.407 € $kg^{-1}$. Assuming that the average production for the year is 120.707 kg $ha^{-1}$, the total risk for temporary fall is 4,320 € $ha^{-1}$.

**Table 3.** Expected value for temporary drop in prices

| Range | Central point (a) | μ | Prob (b) | a·b |
|---|---|---|---|---|
| 0.00–0.15 | 0.050 | 0.000 | 0.000 | 0.000 |
| 0.00–0.30 | 0.150 | 0.000 | 0.000 | 0.000 |
| 0.15–0.45 | 0.300 | 0.161 | 0.341 | 0.102 |
| 0.30–0.60 | 0.450 | 0.286 | 0.606 | 0.273 |
| 0.45–0.75 | 0.600 | 0.025 | 0.054 | 0.032 |
| 0.60–0.90 | 0.750 | 0.000 | 0.000 | 0.000 |
| | Sum | 0.47 | Expected value | 0.407 |

### 3.2 Probability of Permanent Drop in the Price of Peppers

The previous section gives an analysis of the risk of a fall in prices from the mean value to profitability threshold; that is to say, to the value that covers only the variable production costs (0.378 € $kg^{-1}$). However, if the price drops below this figure, the grower will abandon production, in that case the loss is considered definitive. The procedure used is similar to the one in the first section. In this case, only the membership functions that fall below the threshold have been considered. In a similar way to the temporary fall, the expected value obtained has been 0.225 € $kg^{-1}$ which means a loss of 0.153 € $kg^{-1}$ from the threshold (0.378 € $kg^{-1}$). Assuming that the average production is 120.707 kg $ha^{-1}$, the loss is equal to 3,360 € $kg^{-1}$. It should be remembered that this value will be applied to the NPV for the remaining years until the end of the installation's useful life. To do this, it will be necessary to obtain a previous NPV according to the expected net yields until the end of the greenhouse's useful life.

### 3.3 Risk Due to Nematodes and Aphids

Farmers were asked about the possibility that his fields were infected by nematodes or aphids. The loss for not having treated the plants or because the treatment was unsuccessful has been calculated, according to the consulted bibliography [3–8] in 34% for nematodes and 28% for aphids. The probability of occurrence in each field was obtained from experts in the field, answering a survey whose results are shown in Table 4. The product of the loss by the probability and by the total production give a risk of 3,517 € $ha^{-1}$ for nematodes and 215 € $ha^{-1}$ for aphids.

**Table 4.** Nematodes and aphids risk

| Nematodes | | | | | | | | Aphids | | | | | | | |
|---|---|---|---|---|---|---|---|---|---|---|---|---|---|---|---|
| Range | (1) | (2) | (3) | (4) | (5) | (6) | μ | Range | (1) | (2) | (3) | (4) | (5) | (6) | μ |
| 0–5 | 0 | 1 | 1 | 3 | 0 | 0 | 0.48 | 0–0.5 | 3 | 2 | 0 | 0 | 0 | 0 | 0.08 |
| 0–10 | 2 | 1 | 2 | 0 | 0 | 0 | 0.20 | 0–1 | 0 | 2 | 3 | 0 | 0 | 0 | 0.32 |
| 5–15 | 0 | 0 | 2 | 3 | 0 | 0 | 0.52 | 0.5–1.5 | 1 | 0 | 2 | 2 | 0 | 0 | 0.40 |
| 10–20 | 0 | 2 | 2 | 1 | 0 | 0 | 0.36 | 1–2 | 0 | 0 | 0 | 3 | 2 | 0 | 0.68 |
| | | | | | | Prob | 7.44 | | | | | | | Prob | 1.07 |

### 3.4 Risk Due to Bad Quality Irrigation Waters

Loss due to salinity has been obtained from the consulted biography [9–11]. Three levels of salinity in water have been determined, with each one permitting a different production and consequently a loss, which becomes greater according to the increase in salinity of the water. Several studies establish three levels of salinity; for example, Cámara-Zapata et al. [23] carried out a study on the profitability of tomatoes. Given that the quality of the water is not an option for the grower, an occurrence probability for each of the scenarios is considered following the information gathered from the growers. Table 5 summarizes the information provided by the experts. Last column shows the possibility of occurrence for each kind of water. The risk for salinity water is 6.5% of the income.

**Table 5.** Risk due to salinity water

| Stage | (1) | (2) | (3) | (4) | (5) | (6) | μ | Prob | Risk |
|---|---|---|---|---|---|---|---|---|---|
| Top quality | 0 | 0 | 1 | 2 | 2 | 0 | 0.64 | 0.43 | 0.00 |
| 2.5 $dSm^{-1}$ | 0 | 1 | 3 | 1 | 0 | 1 | 0.5 | 0.33 | 0.05 |
| 3.5 $dSm^{-1}$ | 0 | 3 | 1 | 1 | 0 | 0 | 0.32 | 0.21 | 0.19 |
| 6.0 $dSm^{-1}$ | 4 | 1 | 0 | 0 | 0 | 0 | 0.04 | 0.03 | 0.32 |
| | | | | | Average | | 1,84 | | |

### 3.5 Risk Due to Increase in Costs

Next, the risk of whitewashing plastic and drip irrigation has been obtained (see Table 6). According to the answers of the experts, the expected increase has been 4.48. 8.24 y 2.5%, which means 67 € every year for whitewashing (every three years the cost will be 134 since it has estimated an additional whitewashing because of the raining), 618 € for renovating plastic and 115 € for renovating drip irrigation. For the rest of the costs, we have considered an evolution similar to the consumer price index as well as for future income.

**Table 6.** Risk due to increase in costs

| Whitewashing | | | | | | | | Plastic | | | | | | | |
|---|---|---|---|---|---|---|---|---|---|---|---|---|---|---|---|
| Range | (1) | (2) | (3) | (4) | (5) | (6) | μ | Range | (1) | (2) | (3) | (4) | (5) | (6) | μ |
| 0 | 0 | 0 | 1 | 0 | 4 | 0 | 0.72 | 0 | 0 | 3 | 2 | 0 | 0 | 0 | 0.28 |
| 10 | 0 | 2 | 2 | 1 | 0 | 0 | 0.36 | 10 | 1 | 2 | 2 | 0 | 0 | 0 | 0.24 |
| 20 | 3 | 2 | 0 | 0 | 0 | 0 | 0.08 | 20 | 2 | 2 | 1 | 0 | 0 | 0 | 0.16 |
| 30 | 5 | 0 | 0 | 0 | 0 | 0 | 0.00 | 30 | 5 | 0 | 0 | 0 | 0 | 0 | 0.00 |
| | | | | | Increase | | 4.48 | | | | | | Increase | | 8.24 |
| Drip irrigation | | | | | | | | | | | | | | | |
| Range | (1) | (2) | (3) | (4) | (5) | (6) | μ | | | | | | | | |
| 0 | 0 | 0 | 0 | 4 | 1 | 0 | 0.64 | | | | | | | | |
| 10 | 3 | 1 | 1 | 0 | 0 | 0 | 0.12 | | | | | | | | |
| 20 | 4 | 1 | 0 | 0 | 0 | 0 | 0.04 | | | | | | | | |
| 30 | 5 | 0 | 0 | 0 | 0 | 0 | 0.00 | | | | | | | | |
| | | | | | Increase | | 2.50 | | | | | | | | |

### 3.6 Obtaining Fuzzy Decoupled Net Present Value

The annual income is EUR 71,420. The main figures of the revenues and cost are shown in Table 7. Risk costs have also been presented in this table. The net present value for a discount rate of 7.97% is EUR 125,265 and the fuzzy net present value is 72,126 € $kg^{-1}$.

The total sum of risks is 14,414 € $ha^{-1}$ for year 1. Traditional calculation involves determining some annual yields of EUR 25,814, but with this methodology we are able to identify the crop risks, focusing predominantly on problems related to pests and variations in prices. Therefore, the risk-free net yield is reduced to EUR 11,400 in the first year and EUR 13,026 in the latest year.

It should be noted that the percentage of lost income caused by a temporary reduction means a reduction in the same proportion of income for the period, but the percentages for permanent loss means a reduction in NPV for the period, having a greater effect if this reduction occurs in the first years than if it happens in the last years.

NPV has been obtained from estimated annual income and expenses, as well as the correct discounting of structure costs (installation and assembly, plastics, installation of drip irrigation, which are shown in Table 7). Under FDNPV, risks are considered a lower income or higher expense according to cases. As a result, a FDPNV of EUR 72,126 is obtained, using a risk-free interest of 1.62%. By using a risk premium of

**Table 7.** Costs, revenues, risk of increase in costs, net present value (NPV) and fuzzy decoupled net present value (FDNPV).

| Costs | 45,606 | Structure costs | |
|---|---|---|---|
| 1. Variables costs | 34,896 | Structure greenhouse (24) | 103,895 |
| 1.1 Raw materials | 20,051 | Installation of drip irrigation (10) | 4,600 |
| Irrigation water | 1,722 | Plastic covering (3) | 7,500 |
| Seed (Herminio variety) | 5,000 | Additional whitewashing (3) | 1,500 |
| Seedbed | 925 | | |
| Agrocelhone disinfectant | 4,495 | Calculated risks | |
| Pesticides | 2,640 | Loss temporary fall in prices | 4,320 |
| Auxiliary insects | 2,750 | Loss definitive fall in prices | 3,360 |
| Manure | 1,200 | Nematodes | |
| Compost | 1,319 | Aphids | 215 |
| 1.2. Labor | 13,145 | Salinity | 4,640 |
| 1.3. Machinery variable costs | 1,700 | Whitewashing | 67 |
| 2. Machinery fixed costs | 2,680 | Plastic | 618 |
| 3. Other expenses | 8,030 | Drip irrigation (10) | 115 |
| 3.1. Social security | 3,090 | | |
| 3.2. Tax and administrative expenses | 3,440 | Production | 120,707 |
| 3.3. Whitewashing | 1,500 | Income | 71,420 |
| | | Unitary costs (€ $kg^{-1}$) | 0.5917 |
| NPV(7,97%) | 125,265 | FNPV | 72,126 |

6.35% as performed by different authors such as Siegel [24] among others, an NPV of EUR 125,265 is obtained. The result obtained for this particular situation is similar to Lopez-Marin et al. [25].

These calculations illustrate how the use of NPV generates a lot more uncertainty; in this case, the risks to crops or to installations with useful lives that extend over time are introduced into the model thereby increasing the discount rate.

## 4 Conclusions

The methodology used has enabled the isolation and valuation of real risks in the cultivation of greenhouse peppers. Firstly, the risks have been identified concerning the prices of products that experience falls, which can stop the grower from achieving the expected profitability. These are calculated as a risk of a temporary fall in prices and a risk of a permanent fall in prices. If the fall in prices is not very sharp, the grower's profits will be reduced, but it will not change the decision to grow peppers in future campaigns. However, if there is a sharp price fall, the grower will perceive it as a change in trends, and they may replant the crop in the following years. Considering that each one has a particular probability for each risk, a group of expert were required to express their opinions. Fuzzy mathematic was used to deal with the subjective opinions provided by the experts. As a result, the risk for the former has been valued at 4,320 € $ha^{-1}$ $year^{-1}$ per year and the latter at 3,360 € $ha^{-1}$ for the first year, a risk which decreases as the greenhouse's end of useful life approaches. Other risks considered are the loss caused by pests, valued at −2,026 € $ha^{-1}$ per year and risks from loss in water quality (−4,640 € $ha^{-1}$ per year). Finally, although the amount is small, risks from the cost increase in structure elements that need to be replaced have also been considered.

Concerning the comparison between methodologies, it should be noted that the use of FDNPV is much more accurate because it enables quantification of the risks while NPV is limited to increasing the discounting rate. The use of fuzzy logic improves the quality of the risk estimation considering not only historical data available, but the opinion of the experts in the region of cultivation.

## References


1. FAOSTAT. http://www.fao.org/faostat/es/#home. Accessed 21 Jan 2021
2. Magrama. http://www.magrama.gob.es/es/estadistica/temas/estadisticas-alimentacion/observatorio-precios. Accessed 02 Feb 2021
3. Núñez-Zofío, M., Larregla, S., Garbisu, C., Lacasa, A.: Application of sugar beet vinasse followed by solarization reduces the incidence of Meloidogyne incognita in pepper crops while improving soil quality. Phytoparasitica **41**, 181–191 (2013)
4. Ros, M., Garcia, C., Hernandez, T., Lacasa, A., Fernández, P., Fernández, J.A.: Effects of biosolarization as methyl bromide alternative for Meloidogyne incognita control on quality of soil under pepper. Biol. Fertil. Soils **45**, 37 (2008)
5. Sogut, M.A., Elekcioglu, I.H.: Methyl bromide alternatives for controlling Meloidogyne incognita in pepper cultivars in the Eastern Mediterranean Region of Turkey. Turk. J. Agric. For. **31**, 31–40 (2007)
6. Tzortzakakis, E.A., Petsas, S.E.: Investigation of alternatives to methyl bromide for management of Meloidogyne javanica on greenhouse grown tomato. Pest. Manage. Sci. **59**, 1311–1320 (2003)
7. Frantz, J.A., Gardner, J., Hoffmann, M.P., Jahn, M.M.: Greenhouse screening of Capsicum accessions for resistance to green peach aphids (Myzus persciae). HortScience **39**, 1332–1335 (2004)
8. Herman, M.A.B., Nault, B.A., Smart, C.D.: Effects of plant growth-promoting rhizobacteria on bell pepper production and green peach aphid infestations in New York. Crop Prot. **27**, 996–1002 (2008)
9. Aktas, H., Abak, K., Cakmak, I.: Genotypic variation in the response of pepper to salinity. Sci. Hortic. **110**, 260–266 (2006)
10. Anjum, S.A., Farooq, M., Xie, X.Y., Liu, X.J., Ijaz, M.F.: Antioxidant defense system and proline accumulation enables hot pepper to perform better under drought. Sci. Hortic. **140**, 66–73 (2012)
11. Arrowsmith, S., Egan, T.P., Meekins, J.F., Powers, D., Metcalfe, M.: Effects of salt stress on capsaicin content, growth, and fluorescence in a Jalapeno cultivar of Capsicum annuum (Solanaceae). Bios **83**, 1–7 (2012)
12. Fearon, J., Asare, J., Okran, E.O.: Contemporary price trends and their economic significance in the Ashanti region of Ghana. Biol. Agric. Healthc. **4**, 38–47 (2014)

13. Lopez-Marin, J., Brotons-Martinez, J.M., Galvez, A., Porras, I.: Pepper grafting (Capsicum annuum): benefits and profitability. ITEA **112**, 127–146 (2016)
14. Brotons, J.M.: Supuestos de Valoración de Inversiones: Métodos Clásicos y Opciones Reales. Universidad Miguel Hernández, Elche, Spain (2017)
15. Almansa, C., Martínez-Paz, J.M.: What weight should be assigned to future environmental impacts? A probabilistic cost benefit analysis using recent advances on discounting. Sci. Total Environ. **409**, 1305–1314 (2011)
16. Espinoza, R.D., Morris, J.W.F.: Decouple NPV: a simple method to improve valuation of infrastructure investments. Constr. Manage. Econ. **31**, 471–496 (2013)
17. Singh, H., Dunn, B.L., Payton, M., Brandenberger, L.: Selection of fertilizer and cultivar of sweet pepper and eggplant for hydroponic production. Agronomy **9**, 433 (2019)
18. Brotons, J.M., Manera, J., Conesa, A., Porras, I.: A fuzzy approach to the loss of green colour in lemon (Citrus lemon L. Burm. f.) rind during ripening. Comput. Electron. Agric. **98**, 222–232 (2013). https://doi.org/10.1016/j.compag.2013.08.011
19. CARM. Consejería de Agricultura de la Región de Murcia. https://caamext.carm.es/esamweb/faces/vista/seleccionPrecios.jsp. Accessed 20 Mar 2021
20. Rameshwaran, P., Tepe, A., Yazar, A., Ragab, R.: The effect of saline irrigation water on the yield of pepper: experimental and modelling. Irrig. Drain. **64**, 41–49 (2015)
21. Savvas, D., et al.: Interactions between salinity and irrigation frequency in greenhouse pepper grown in closed-cycle hydroponic systems. Agric. Water Manage. **91**, 102–111 (2007)
22. Stock Exchanges and Markets. https://www.bmerf.es/esp/aspx/comun/posiciones.aspx?Mercado=SDP&menu=47. Accessed 30 Mar 2021
23. Cámara-Zapata, J.M., Brotons-Martínez, J.M., Simón-Grao, S., Martinez-Nicolás, J.J., García-Sánchez, F.: Cost-benefit analysis of tomato in soilless culture systems with saline water under greenhouse conditions. J. Sci. Food Agric. **99**, 5842–5851 (2019)
24. Siegel, J.J.: Perspectives on the equity risk premium. Financ. Analysts J. **61**(6), 61–73 (2005). https://doi.org/10.2469/faj.v61.n6.2772
25. Lopez Marin, J.J., Galvez, A., del Amor, F.M., Brotons, J.M.: The financial valuation risk in pepper production: the use of decoupled net present value. Mathematics **9**(1), 13 (2020). https://doi.org/10.3390/math9010013

# The Effect of Energy Prices on Mexican Households' Consumption

María Guadalupe García Garza, Jeyle Ortiz Rodríguez(✉), and Esteban Picazzo Palencia

Facultad de Contaduría Publica y Administracion, Universidad Autonoma de Nuevo Leon, 66451 San Nicolás de los Garza, NL, Mexico
jeyle.ortizrdr@uanl.edu.mx

**Abstract.** Historically, Mexico has been a relevant actor in the international energy market. In 1984, Mexico was the fourth largest producer of oil in the world. However, since 2000 oil production in Mexico has been seriously affected due to unfavorable national economic and institutional aspects, and adverse international conditions. After being the sixth largest petroleum producer in 2000 (3,460 barrels per day), in 2019, Mexico occupied the fourteenth place in oil production in the world (1,914 barrels per day). Reductions in country's energy production as well as a high domestic demand generated serious economic problems to keep the existing high subsidies in the market. In poor and developing countries, when income increases, households tend to invest in private automobiles rather than using public transportation. As a consequence, in 2013, an energy reform was promulgated in Mexico. Using in-formation from the National Survey of Household Income and Expenditure, this research estimates a Quadratic Almost Demand System (QUAIDS) for 2018. Results indicate that the estimated elasticity prices indicate that the demand for energy is inelastic. When prices increase, households do not tend to change their consumption. In the face of increases in energy prices, households will tend to cut expenditure shares of other goods. Changes in prices have a greater impact on low-income household well-being compared to high-income households.

**Keywords:** Energy prices · Mexican households · Quadratic almost demand system

## 1 Introduction

Historically, Mexico has occupied a significant place in the international energy market. In 1984, Mexico's oil production volume ranked as the fourth in the world. However, since 2000, oil production has declined as a result from an unfavorable international environment as well as economic and institutional aspects. Mexico, as the sixth oil producer in the world in 2000 (3.460 million barrels per day) dropped to the fourteenth place in 2019 (1.914 million barrels per day). In the same way, in 2002 Mexico was the fifth oil exporter (1.956 million barrels per day), but in 2016 dropped to the thirteenth place in the world with 1.236 million barrels per day [1]. The energy sector has been


M. d. P. Rodríguez García et al. (Eds.): XX SIGEF 2021, LNNS 384, pp. 47–56, 2022.
https://doi.org/10.1007/978-3-030-94485-8_4

key for the Mexican economy. According to the National Institute of Statistics and Geography (INEGI, 2019), in 1996 the production value from oil and gas extraction accounted for 9.21% of the Gross Domestic Product (GDP) of Mexico. The share of production value from oil and gas extraction in GDP declined from 8.00% in 2000 to 3.25% in 2019 [2].

The steep decrease in oil production in Mexico can be mainly related to the degradation of natural resources, as well as to the barriers that Petróleos Mexicanos (Pemex, Mexico's national oil company) faced from the rise of deep-water oil exploitation and extraction [3]. Deepwater extraction is expensive and entails a lower probability to succeed; it is a global problem. The conventional oil reserves known as light oil, based on physical characteristics and location, found in fixed platforms or shallow water rigs, even though it is easy to extract, it has started to exhaust their potential; unlike the heavy oil of higher viscosity and found in deep-water. Although hydrocarbons production in deepwater in 2003 barely reached 1% of the total global production, by 2013 accounted for 8%. As for hydrocarbon reserves, estimations from the Mexican Senate of the Republic report that 83% of such reserves by 2012 were significantly dropping or about to drop, which was causing oil production decrease [4].

Moreover, since 2004, after Mexico reached a maximum peak of oil production (3.85 million barrels per day), the annual average decrease has been of 4.82% [1], causing insufficient oil and gas supply, and the need to import diesel, gas, gasoline and petrochemicals, reflecting at the end a deficit in the trade balance of hydrocarbons.

Furthermore, since the country's public finances highly depend on the budget revenue from Pemex operations, that contributed to insufficient levels of investment in technology, infrastructure and exploration projects, hydrocarbons extraction to promote the energy sector and, therefore, the industrial development of the country was affected [4].

Also, since the middle of the twentieth century, Mexican households have experimented social and economic transformations that essentially had an impact in lifestyle. Higher education levels and the participation of women in the labor force, impacted the family economy, causing transformations in the composition and structure of families. New dynamics of population, new lifestyles, more household equipment, and higher incomes led to a higher demand for energy [5]. The decrease in the production and production capacity of energy and electricity, combined with the consumers' high demand, as well as the decrease of oil price, created severe economic problems for the government to maintain high subsidies. Only in 2012, the subsidies in Mexico for the consumption of energy products (electricity, gas and gasoline) accounted for 24,453 million dollars [1]. Such subsidies for energy poor households were intended to promote the industrial activity and also to grant access to energy goods.

As a response to the energy sector scenario, to the high financial burden on public finances and, to the main political project to the end of 2013, the Mexican government initiated the transformation of the energy sector, impacting public and private policies to increase the productivity of the sector and to trigger the economic growth. By 2013 the energy reform is enacted to increase the efficiency and lower the prices under a proper institutional design. The reform mainly achieved the elimination of subsidies to the energy sector, which resulted in more volatile prices of energy and less control at domestic level.

Although the goal of the reform was to increase the productivity in the sector to lower energy prices and promote economic development in Mexico, these changes are expected to occur in the long term. In the short term, energy prices are likely to increase. The effect of changes in prices on households' consumption have been scarcely explored in Mexico. As expected, a drop in subsidies is likely to have negative impacts on households' well-being. As price elasticities of energy demand tend to be inelastic in the short term, poor households are more affected by changes in prices [6]. Using a Quadratic Almost Ideal Demand System (QUAIDS), the objective of this research is to analyze the effect of energy prices on Mexican households' consumption by income level in 2018.

## 2 Theoretical Background

Governments use energy pricing policies (electricity, gas and gasoline) to mitigate the impact of volatile prices in households, prevent inflation, accelerate the economy and achieve social welfare. Many producing countries have adopted this method to promote an equitable distribution of wealth. This situation has prevailed in Latin America using a method to control energy prices, which was intensified after high rises in the international oil price at the beginning of 2000 and lasted until 2014 [7]. Nevertheless, these measures result in a high burden to the public finances of the countries, that in many cases has led to enact energy reforms for the liberalization of prices in the sector. The first consequence of such reforms is the increase of energy products prices.

The evidence shows that the impact of prices in energy consumption is asymmetric [8–14]. As the energy goods prices increase, the consumption decreases, but in less proportion. An important reason for this asymmetric response is that consumers cannot easily replace the more energy-efficient goods once the energy prices begin to decrease from higher levels [6]. In addition, increases in electricity prices indirectly increase the price of other goods and services in an economy. Households are exposed to the impacts of electricity prices depending on how they distribute their money. The increase in electricity prices mainly affects to low-income households in every country. The consumption of electricity in Mexico, as in other countries, is related to changes in housing conditions and equipment, as well as to practices regarding the use of energy goods in households. Mexico experiments socioeconomic inequality, but an increase in the mid income has been recorded. In addition, Mexico has implemented credit expansion and housing conditions that were beneficial for larger segments of the population than in the past years [16]. The use of electricity is significant for the domestic and residential energy consumption and it has increased during the past decades. The residential consumption of electricity rose its share by over three quarters of electricity consumption in Mexico from the last two decades [16].

Regarding gasoline, the variables that establish gasoline consumption are mainly price and number of vehicles and have differentiated impacts per income group in Mexico. According to INEGI, between 1991 and 2012 the number of private vehicles grew from 6.6 million to 22.4 million, an increase of 239%. In addition, the new vehicles efficiency only increased 6% between 1988 and 2008 [17], which, as a whole, resulted in more demand for gasoline. Visibly the transportation is the sector with major increase and regarding fuels, natural gas and gasoline. An increase in the price of gasoline, even when represents a decrease of gasoline demand in all the deciles, has greater impact in the deciles of low income, that is, high-income households are more inelastic to the price of gasoline. The five deciles of low income only account for 16% of the total demand for gasoline. When the price of gasoline changes, the demand tends to be less elastic in sectors that more consume, a policy to increase gasoline price only decrease the consumption in those with low income and few cars [18]. On the other hand, natural gas is used essentially for electricity generation, the energy sector consumption and the industry.

Some studies across countries explore the households' well-being through analyzing consumption patterns of electricity, oil byproducts, natural gas and transportation in different income levels, in rural and urban areas, and if the household can pay such goods and services in case of changes in prices or subsidies [19]. The results show that rising the prices without previously evaluating the impact can affect the households' well-being causing reduction in the consumption of basic goods such as food; health deterioration due to the use of cheap fuel substitutes such as charcoal and wood for cooking; and pollution for using low quality transportation.

# 3 Methodology

## 3.1 Data

The data regarding expenditure of households used in this research was sourced from the National Survey of Household Income and Expenditure (ENIGH) 2018 collected by the National Institute of Statistics and Geography. The ENIGH gathers information on expenditure, characteristics of people living in households and housing aspects. The information on energy prices was obtained from the Energy Regulatory Commission.

## 3.2 Model

For the demand-system estimation of energy goods will be used the Quadratic Almost Ideal Demand System (QUAIDS). This system was developed by Banks, Blundell and Lewbel in 1997 as an extension of the Almost Ideal Demand System (AIDS) for obtaining a better projection of the reality by incorporating more terms related to income and for providing more flexibility to the models. That is, to obtain a linear behavior from Engel curves of some goods in the logarithm of total expenditure and also, to behave as luxury goods in some income levels, and as necessity goods in others.

Banks, Blundell and Lewbel [20] start with the following general form of demands consistent with the empirical evidence of Engel curves:

$$w_i = a_i(p) + b_i(p)\log m + c_i(p)g(m) \tag{1}$$

where:

$w_i$ = total expenditure proportion for good $i$ (food, gas, gasoline, electricity and other goods).

$p$ = prices vector for $n$ goods.
$m$ = total expenditure.
$a_i, b_i, c_i, g(m)$ = differentiable functions.

The above equation shows that the total expenditure proportion for good $i$ has a linear relationship with the total expenditure logarithm, and that the last term allows for the nonlinearity of the Engel curve, which will be equal to zero in the presence of Price-Independent Generalized Logarithmic (PIGLOG) preferences. The Banks, Blundell and Lewbel's demand system [20] is derived from the maximization of the following indirect utility function developed as an extension of the AIDS:

$$\log V = \left\{ \left( \frac{\log m - \log a(p)}{b(p)} \right)^{-1} + \lambda(p) \right\}^{-1} \quad (2)$$

where $V$ is the indirect utility, $m$ the total expenditure, *[log m – log a(p)]/b(p)* the utility function of the AIDS and, with PIGLOG preferences, $\lambda(p)$ is a zero-grade homogeneous and differentiable function that allows for non-linear Engel curves and would be zero in PIGLOG preferences, and *a(p)* and *b(p)* represent the subsistence cost and shadow cost, respectively. Besides, it is assumed that:

$$\log a(p) = \alpha_o + \sum\nolimits_{i=1}^{N} \alpha_i \log p_i + \frac{1}{2} \sum\nolimits_{i=1}^{N} \sum\nolimits_{j=1}^{N} \gamma_{ij} \log p_i \log p_j \quad (3)$$

$$b(p) = \prod\nolimits_{i=1}^{n} p_i^{\beta_i} \quad (4)$$

$$\lambda(p) = \sum\nolimits_{i=1}^{N} \lambda_i \log(p) \quad (5)$$

$$\sum\nolimits_{i=1} \lambda_i = 0 \quad (6)$$

Applying the Roy's theorem to the Eq. (2) it is obtained the total expenditure proportion on each good:

$$w_i = \frac{\partial \log a(p)}{\partial \log p_i} + \frac{\partial \log b(p)}{\partial p_i} \log y + \frac{\partial \lambda}{\partial \log p_i} \frac{1}{b(p)} (\log y)^2 \quad (7)$$

QUAIDS is obtained replacing the parameterization of prices in the expenditure share, where the expenditure share of good $\imath$ for household $h$ is expressed as follows:

$$w_{ih} = \sum\nolimits_{m=1}^{M} \alpha_{im} Z_{mh} + \sum\nolimits_{j=1}^{K} \gamma_{ij} \log p_{jh} + \beta_i \log\left[\frac{X_h}{a(p)}\right] + \frac{\lambda_i}{b(p)} \left\{ \log\left[\frac{X_h}{a(p)}\right] \right\} + \xi_{ih} \quad (8)$$

Also, in order to satisfy homogeneity, and Slutsky symmetry the following conditions are needed:

$$\sum\nolimits_{i=1}^{N} \gamma_{ij} = 0 \quad (9)$$

$$\sum\nolimits_{i=1}^{N} \alpha_i = 1 \quad (10)$$

$$\sum_{i=1}^{N} \beta_i = 0 \tag{11}$$

$$\gamma_{ij} = \gamma_{ji} \tag{12}$$

Given the Eq. (8), it can be seen that income elasticity and price demand on good $i$ for the household $h$ are given by the Eq. (13) and Eq. (14), respectively.

$$e_{ij} = \frac{\mu_{ij}}{W_{ih}} - \delta_{ij} \tag{13}$$

$$\varepsilon_{ih} = \frac{\mu_i}{W_{ih}} + 1 \tag{14}$$

where:

$$\mu_i = \frac{\partial w_i}{\partial 1nm} = \beta_i + \frac{2\lambda_i}{b(p)}\left\{1n\left[\frac{m}{a(p)}\right]\right\} \tag{15}$$

$$\mu_{ij} = \frac{\partial w_i}{\partial 1nP_j} = \gamma_{ij} - \mu_i\left(\alpha_j + \sum_k \gamma_{jk} 1nP_k\right) - \frac{\lambda_i\beta_j}{b(P)}\left\{1n\left[\frac{m}{a(P)}\right]\right\} \tag{16}$$

$$\delta_{ij} = \begin{cases} 1 \ \textit{if} \ i = j \\ 0 \ \textit{if} \ i \neq j \end{cases} \tag{17}$$

## 4 Results

Table 1 shows the estimations results of QUAIDS parameters and, Table 2 shows the demand income elasticities of the households for different categories using QUAIDS parameters estimations. The income elasticity of food is less than one, which shows that food is a necessity good. In the same way, the income elasticity of gas and electricity is less than the unit. On the other hand, the income elasticity of other goods and gasoline are greater than one.

Table 3 shows the price elasticity of demand for five categories by households' income level, as well as the cross-price elasticity. Elasticities of all categories show a negative sign, which implies that the demand of goods has a negative slope, that is, if the price of the good increases, the demand decreases. The price elasticity of food and electricity is inelastic for all households independently of the income level. The demand for gas and gasoline is elastic for low and middle-income households; however, for high-income households, the price elasticity of such goods categories is inelastic. Regarding other goods, price elasticity is close to the unit, but the more increases the income level of households, the more inelastic is the price.

The cross-price elasticity for food demand with electricity and gas is negative and significant for all households, which indicates that are complementary goods. The cross-price elasticity of electricity with gas and gasoline is not significant, which indicates that changes in gas or gasoline prices do not affect the demanded amount for electricity. Although when considering all households, changes in gasoline price do not affect food

**Table 1.** Estimation of QUAIDS parameters

| Parameters | Category | Coefficient | Standard error | z |
|---|---|---|---|---|
| α (alpha) | Food | 0.79612 | 0.2072 | 3.84 |
| | Gasoline | 0.32278 | 0.0856 | 3.77 |
| | Gas | 0.18195 | 0.0271 | 6.70 |
| | Electricity | 0.65314 | 0.1381 | 4.73 |
| | Other goods | 0.35167 | 0.1350 | 2.61 |
| β (beta) | Food | −0.03373 | 0.0098 | −3.44 |
| | Gasoline | −0.02170 | 0.0041 | −5.25 |
| | Gas | −0.05179 | 0.0083 | −6.25 |
| | Electricity | −0.04092 | 0.0066 | −6.19 |
| | Other goods | 0.01633 | 0.0055 | 2.99 |
| γ (gamma) | Food – food | 0.03830 | 0.0047 | 8.21 |
| | Gasoline – food | −0.00187 | 0.0015 | −1.24 |
| | Gas – food | −0.01114 | 0.0024 | −4.60 |
| | Electricity – food | −0.02165 | 0.0021 | −10.12 |
| | Other goods – food | −0.01133 | 0.0011 | −10.16 |
| | Gasoline – gasoline | 0.01553 | 0.0072 | 2.15 |
| | Gas – gasoline | −0.00071 | 0.0010 | −0.71 |
| | Electricity – Gasoline | −0.00030 | 0.0009 | −0.32 |
| | Other goods – Gasoline | −0.00257 | 0.0005 | −5.34 |
| | Gas- Gas | 0.01495 | 0.0023 | 6.43 |
| | Electricity – Gas | 0.00437 | 0.0018 | 2.47 |
| | Other goods – Gas | −0.00788 | 0.0009 | −9.16 |
| | Other goods – other goods | 0.02513 | 0.0091 | 2.75 |
| λ (lambda) | Foods | −0.00688 | 0.0012 | −5.63 |
| | Gasoline | 0.00099 | 0.0003 | 3.03 |
| | Gas | 0.00013 | 0.0001 | 2.32 |
| | Electricity | −0.00401 | 0.0008 | −4.85 |
| | Other goods | 0.00228 | 0.0007 | 3.31 |

Note: Table created using data from ENIGH, 2018.

**Table 2.** Income elasticity of households' demand by income level

| Category | All households | Low income | Middle income | High income |
|---|---|---|---|---|
| Food | 0.728 | 0.743 | 0.791 | 0.693 |
| Gasoline | 1.084 | 1.118 | 1.153 | 1.149 |
| Gas | 0.842 | 0.817 | 0.886 | 0.893 |
| Electricity | 0.851 | 0.826 | 0.925 | 0.902 |
| Other goods | 1.464 | 1.591 | 1.435 | 1.552 |

Note: Table created using data from ENIGH, 2018.

**Table 3.** Price elasticity of demand for five categories by households' income level

| | | Food | Gasoline | Gas | Electricity | Other goods |
|---|---|---|---|---|---|---|
| All households | Food | −0.324* | 0.045 | −0.062* | −1.050* | 0.462* |
| | Gasoline | 0.685 | −1.117* | 1.232 | −0.567 | 0.099* |
| | Gas | −0.064* | 0.016 | −1.236* | 0.052 | 0.148* |
| | Electricity | −1.256* | −0.556 | 0.029 | −0.422* | 1.653* |
| | Other goods | 0.365* | 0.087* | 0.165* | 0.894* | −0.934* |
| Low-income | Food | −0.321* | 0.045* | −0.060* | −1.004* | 0.491* |
| | Gasoline | 0.723* | −1.095* | 1.657 | −0.380 | 0.096* |
| | Gas | −0.062* | 0.022 | −1.162* | −0.049 | 0.153* |
| | Electricity | −1.202* | −0.296 | −0.044 | −0.392* | 1.719* |
| | Other goods | 0.336* | 0.091 | 0.226* | 1.102* | −0.710* |
| Medium-income | Food | −0.314* | 0.048 | −0.055* | −0.701* | 0.468* |
| | Gasoline | 0.464 | −1.072* | 1.714 | −0.532 | 0.046* |
| | Gas | −0.062* | 0.006 | −1.149* | −0.061 | 0.081* |
| | Electricity | −1.236* | −0.467 | −0.054 | −0.399* | 1.361* |
| | Other goods | 0.318* | 0.041* | 0.120* | 0.873* | −0.838* |
| High-income | Food | −0.327* | 0.075 | −0.062* | −1.239* | 0.403* |
| | Gasoline | 0.557 | −0.860* | 1.773 | −0.556 | 0.102* |
| | Gas | −0.067* | 0.008 | −0.964* | 0.040 | 0.117* |
| | Electricity | −1.340* | −0.722 | 0.013 | −0.430* | 1.354* |
| | Other goods | 0.264* | 0.058* | 0.173* | 0.888* | −1.098* |

*Note*: Table created using data from ENIGH, 2018.
*p < 0.001.

demand, for low-income households it is significant and positive, which indicates that gasoline and food are substitute goods for low-income households.

The cross-price elasticity between other goods and electricity indicates that these goods are substitutes for all-income levels households, but the scale of such elasticity is slightly greater for low-income households.

## 5 Conclusions and Discussion

This research analyzes how changes in energy goods prices impacted households' consumption in Mexico in 2018. The QUAIDS model and data of ENIGH were used to estimate income, price and cross-price elasticities of demand.

The estimated price elasticity indicates that the demand for energy is inelastic. When prices increase, households do not tend to change their consumption. In the face of increases in energy prices, households will tend to cut expenditure shares of other goods. Changes in prices have a greater impact on low-income households well-being compared to high-income households.

Although the price elasticity of the demand is inelastic for all households, low-income households are more impacted. The results of this research indicate that increases in electricity prices not only result in less electricity, but also less food, which entails more vulnerability. These results are also observed when gas prices increase. Impacts identified by this research agree with the ones found in other studies [21, 22].

Changes in electricity and gas prices not only result in reduced consumption of these goods by low-income households, but also reduce the expenditure on other basic goods such as food and, leads to the possible use of low-cost fuel such as charcoal and wood as substitute goods for cooking, which pose a health risk for the people living in households.

The expected results from the energy reform have not yet led to reduce energy prices, mainly due to unfavorable situations in the international market. In this scenario low-income households are the most vulnerable to volatility of energy prices. High-income households have been able to choose alternatives like solar panels, energy-efficient appliances and hybrid cars as energy consumption substitutes. This research shows that electricity consumption subsidies for low-income households have not been enough to compensate the volatility of electricity prices. The challenge for future research is comparing energy price elasticities in different periods and measuring the energy reform impact within a broader time frame, as well as identifying complementary support for lower-income households to face the volatility of energy prices without declines in well-being.

## References

1. Energy Information Administration U.S. (EIA). https://www.eia.gov/international/rankings/country/MEX/. Accessed 31 Mar 2020
2. Instituto Nacional de Estadística Geografía (INEGI). https://www.inegi.org.mx/temas/ingresoshog/. Accessed 01 Apr 2020
3. Del Río, J., Rosales, M., Ortega, V., Maya, S.: Análisis de la reforma energética. Instituto Belisario Domínguez, Senado de la República, Ciudad de México (2016)

4. Senado de la República: Dictamen de las Comisiones Unidas de Puntos Constitucionales; de Energía, y Estudios Legislativos, Primera, con proyecto de decreto por el que se reforman y adicionan los artículos 25, 7 y 28 de la Constitución Política de los Estados Unidos Mexicanos en materia de energía, Senado de la República, Ciudad de México (2013)
5. Sánchez, P.: Hogares y consumo energético. Revista Digital Universitaria de la UNAM **13**(10), 1–8 (2012)
6. Huntington, H., Barrios, J., Arora, V.: Review of key international demand elasticities for major industrializing economies. Energy Policy **133**, 110878 (2019)
7. Feng, K., Hubacek, K., Liu, Y., Marchán, E., Vogt-Schilb, A.: Efectos distributivos de los impuestos a la energía y de la eliminación de los subsidios energéticos en América Latina y el Caribe, Inter-American Development Bank Working Paper Series No. IDB-WP-947 (2018)
8. Dargay, J.: The irreversible effects of high oil prices: empirical evidence for the demand for motor fuels in France, Germany and the UK. In: Hawdon, D. (ed.) Energy Demand: Evidence and Expectations, pp. 165–182. Surrey University Press, Guildford, UK (1992)
9. Dargay, J., Gately, D.: The imperfect price-reversibility of non-transport oil demand in the OECD. Energy Econ. **17**(1), 59–71 (1995)
10. Dargay, J., Gately, D.: The demand for transportation fuels: imperfect price reversibility? Transport. Res. B: Methodol. **31**(1), 71–82 (1997)
11. Dargay, J., Gately, D.: World oil demand's shift toward faster growing and less price-responsive products and regions. Energy Policy **38**(10), 6261–6277 (2010)
12. Huntington, H.G.: Short- and long-run adjustments in U.S. petroleum consumption. Energy Econ. **32**(1), 63–72 (2010)
13. Gately, D., Huntington, H.G.: The asymmetric effects of changes in price and income on energy and oil demand. Energy J. **23**(1), 19–55 (2002)
14. Walker, I., Wirl, F.: Irreversible price-induced efficiency improvements: theory and empirical application to road transportation. Energy J. **14**(4), 183–205 (1993)
15. Wolfram, C., Shelef, O., Gertler, P.: How will energy demand develop in the developing world? J. Econ. Perspect. **26**(1), 119–138 (2012)
16. Secretaría de Energía, información estadística. http://sie.energia.gob.mx. Accessed 01 June 2021
17. Sheinbaum-Pardo, C., Chávez-Baheza, C.: Fuel economy of new passenger cars in Mexico: trends from 1988 to 2008 and prospects. Energy Policy **39**(12), 8153–8162 (2011)
18. Sánchez, A., Islas, S., Sheinbaum, C.: Demanda de gasolina y la heterogeneidad en los ingresos de los hogares en México. Investigación Económica **74**(291), 117–143 (2015)
19. Bacon, R., Bhattacharya, S., Kojima, M.: Expenditure of low-income households on energy. Evidence from Africa and Asia. Oil, Gas, and Mining Policy Division Working Paper. World Bank (2010)
20. Banks, J., Blundell, R., Lewbel, A.: Quadratic Engel curves and consumer demand. Rev. Econ. Stat. **79**(4), 527–539 (1997)
21. Moshiri, S., Martínez, M.A.: The welfare effects of energy price changes due to energy market reform in Mexico. Energy Policy **113**, 663–672 (2018)
22. Ortega, A., Medlock, K.B.: Price elasticity of demand for fuels by income level in Mexican households. Energy Policy **151**(May 2020), 112132 (2021)

# Management and Accounting

# How to Cope with Complexity in Decision-Making: An Application of Fuzzy Qualitative Comparative Analysis in the Triage Process

Lorella Cannavacciuolo[1], Cristina Ponsiglione[1], Simonetta Primario[1], Ivana Quinto[1](✉), Maria Teresa Iannuzzo[2], and Giovanna Pentella[2]

[1] Department of Industrial Engineering, University of Naples Federico II, Piazzale Tecchio, 80, 80125 Naples, Italy
ivana.quinto@unina.it

[2] Buon Consiglio Fatebenefratelli Hospital in Naples, Via Alessandro Manzoni, 80123 Naples, Italy

**Abstract.** Qualitative Comparative analysis (QCA) is a method used to test theory-based conditions considering multiple interrelated variables that lead to the same outcome. QCA has been applied in several fields (political, sociological, organizational, and marketing), but recently studies are bridging configurational analysis using fsQCA with complexity theory in sub-disciplines of business and management. This paper aims at highlighting the usefulness of QCA also in decision-making. More specifically, we show how the QCA allow researchers to figure out how the interaction between different factors (individual, organizational, environmental) affect the cognitive heuristic process. The cognitive heuristic process is a shortcut strategy used by an individual to take a decision leveraged on limited information, time and processing capacity. We consider the cognitive heuristic process of Triage, performed by nurses in the Emergency Department. The Triage process requires a complex interaction between nurses and patients within a more or less turbulent environment: verbal information (the patient's history), visual cues (non-verbal communication), and possibly vital signs, determine the outcome of the decision-making process. Contextual factors (i.e. organizational rules and environmental constraints) and individual nurse's experience, knowledge, and intuition can cause the variation of nurses' decision-making in assessing the urgency of any individual patient to receive the care. In addition, we propose some practical implications to improve the Triage process based on the results of the QCA application.

**Keywords:** Triage · Fuzzy qualitative comparative analysis · Fuzzy QCA · Complexity · Decision-making process

## 1 Introduction

Qualitative comparative analysis (QCA) is a method aimed at understanding the complexity of a phenomenon, focusing on a set of theory-based conditions rather than considering separately the effects of individual variables.


M. d. P. Rodríguez García et al. (Eds.): XX SIGEF 2021, LNNS 384, pp. 59–72, 2022.
https://doi.org/10.1007/978-3-030-94485-8_5

QCA is used in different research fields and also business and management scholars are showing an increasing interest towards this method. Kraus et al. [1] offer an overview on the existing research on QCA in different management research areas, highlighting its relevance for a deeper understanding of a complex phenomenon.

The QCA presents three different versions based on analytical and software tools [2, 3]: the crisp set (csQCA) version, the version based on fuzzy sets (fsQCA), the multi-value version (mvQCA).

More in depth, QCA is a comparative case-oriented technique that allows to perform an analysis on a small or intermediate number of empirical cases in order to identify the configurations of causally relevant conditions linked to the outcome under investigation [4]. In fact, this method analyzes the relationships between the conditions and identify the configurations of conditions than can produce a given outcome. It is an inductive research method inspired by complexity theory based on the principles of conjunction, equifinality, and causal asymmetry [5]. Conjunction concerns the evidence that we can obtain an outcome with different configurations of conditions. Equifinality means that we can find more effective configurations related to a given output, each configuration is sufficient for reaching the output. Causal asymmetry implies that some conditions can be absent in a configuration or inversely related and present in another one. This situation is due to the fact that "the directionality between X and Y depends on what additional simple conditions occur in given contexts" [6], where X is the condition and Y is the output.

QCA methodology seems to be the answer for understanding phenomena for which it is not the single variable but the relationships between them that determine a result. This feature of QCA is also well suited in the field of decision-making based on cognitive heuristics. Cognitive heuristics represent cognitive shortcut strategies used to take a decision leveraged on limited information, time and processing capacity [7]. Decision-makers apply cognitive heuristics to look for the solution to a problem: his/her strategy is based on a selection of information leading him/her to simplify the process of elaboration. The Basel and Bruhl's [8] review highlights the relevance of the cognitive heuristic strategy in the management field: cognitive heuristic become a suitable strategy to take decisions in uncertain and complex environments. The management perspective recognizes that the cognitive heuristic strategy is influenced not only by the individual characteristics of decision-maker but also by organizational and environmental factors. Many scholars have highlighted the influence of organizational factors [9–12], but an in-depth analysis of which factors, between individual and organizational ones, prevail in the decision-making process remains an area that has not yet been explored.

Against this backdrop, the aim of this research is to show how QCA can be the answer to better analyse the heuristic decision-making, identifying how the interaction of organizational, environment and individual factors affect the decision-maker answering to specific situations. We apply the fsQCA to the Triage process performed by nurses in an Emergency Department. The fuzzy-set-based variant is used to consider the granularity of information and data collected during the fieldwork. The possibility to use both fuzzy variables and crisp variables is another reason that makes this method well suited for the context of this study.

The aim of Triage is to prioritize the patients arriving at the Emergency Department according to their urgency to receive care. The nurses' decision-making relies on cognitive heuristics [13]. In fact, nurses attribute patient's priority level according to his/her experience under uncertain conditions of his/her everyday work through an interpretation process of patient's clinical conditions based on few cues and small laps of time [14, 15]. The Triage process requires a complex interaction between nurses and patients within a more or less turbulent environment [16]: verbal information (the patient's history), visual cues (non-verbal communication), and possibly vital signs, determine the outcome of the decision-making process. Contextual factors (i.e. organizational rules and environmental constraints) and individual nurse's experience, knowledge, and intuition can cause the variation of nurses' decision-making in assessing the urgency of any individual patient [16–18].

Based on these premises, we show the effectiveness of fsQCA method for a deeper understanding of the heuristic cognitive decision-making of the Triage. More in detail, the application of fsQCA allows us to answer to the following research question: "What are the combinations of context and individual factors that affect positively the quality of the Triage process?".

The contribution of this research resides not only in the methodological perspective (i.e. fsQCA as prominent method for cognitive heuristic decision-making debate) but also in the practical implications. Even before Covid-19, scholars have given great attention to make Triage more efficient to streamline the processing of patients and reduce the waiting time of more urgent patients [19–22]. Few studies investigate nurses' decision-making quality even if understanding what factors influence decisions is a relevant aspect to improve Triage effectiveness [23]. These studies evaluate the quality of decisions based on scenarios or on patient's records retrospectively collected. Furthermore, they only assess the accuracy. Unlike them, our research is not limited to evaluating the accuracy but highlights the "configuration of conditions" that allowed the achievement of the decision accuracy. In addition, our research is not based on scenarios or patient's records retrospectively collected and analyzed, but our data were gathered directly through the observation of the Triage process and of nurses assigning the priority codes. In a practical perspective, this study can also offer interesting insights on how to sustain the improvement of the Triage process.

## 2 QCA Methodology

Qualitative Comparative Analysis investigates how a combination of conditions produces a given outcome, considering a relatively small or medium number of empirical cases and using principles of Boolean algebra [2, 24–26].

QCA conceptualizes the conditions as sets, wherein if the condition is not present the case has a membership degree equal to 0 for this condition otherwise the membership is equal to 1 if the condition is fully present. In fsQCA the membership is a value between 0 and 1, as the condition can be present but not fully present. QCA assesses the necessity and e the sufficiency of the relationships between conditions and the outcome. We have a necessary situation if the condition has to be present for the outcome to occur. In a sufficient situation, the condition can produce the outcome and it is a subset of the outcome itself.

In this chapter, we adopt a fsQCA approach to analyze row data obtained from the analyzed and observed real cases, according to the following steps:

1. Calibration, aiming at transforming data into fuzzy sets.

   The cases are represented as configuration of conditions through a matrix whose rows are cases and whose columns are the conditions and the outcome. The calibration process concerns the assignment of set membership scores to cases using theoretical and empirical evidence [26]. The membership value can be equal to: i) 0 (the condition is fully outside the set); ii) a value ranging from 0 and 1 (the condition belongs to the case with a certain membership); iii) 1 (the condition is fully inside the set).

   This is the critical step of fsQCA as the researcher has to define both the conditions and the fuzzy membership of conditions to the cases. The conditions are the relevant variables inferred by the observation or the literature analysis about the investigated problem. The fuzzy membership of the variables derived by the theoretical guidance and by the strong experience on the empirical cases developed by researchers. The calibration was assisted by the R package for fsQCA.
2. Constructing the truth table, aiming at obtaining a distribution of the cases across all possible configurations.

   The truth-table groups empirical cases based on the fact that they show the presence or absence of the outcome. The truth-table shows as many rows as there are combinations of causal conditions, obtaining a matrix of 2k rows, where k is the number of causal conditions. Based on the truth table, we assess the consistence and coverage of configurations. The consistence measures whether a configuration produces or not the outcome in real data [27]; it ranges from 0 to 1, with 1 indicating perfect consistency. The coverage evaluates how many cases are covered by a given set of conditions and it also ranges from 0 to 1. In many research domains, consistency and coverage of a subset relation can be contrasting measures and the researcher has to set a trade-off between the two, based on the object of investigation, number of causal conditions, and available cases [28].
3. Identification and interpretation of consistent and empirically relevant patterns (causal configurations of conditions) pertaining to the outcome.

   The third stage of the analysis simplifies combinations and minimizes solutions, reducing the number of configurations of the initial truth table by specifying the consistency threshold [5, 25, 29, 30]. Researchers analyse the truth-table to figure out connections between configurations of causal conditions and the outcome. The causal conditions can be necessary for an outcome if instances of the outcome constitute a subset of the instances of the causal condition, whereas if the instances of the causal condition constitute a subset of the outcome it is sufficient.

Most of the steps described above are taken with the help of software specifically developed in the context of QCA research. In this study, the package fsQCA 3.0 is adopted.

## 3 Case Design

Field research has been conducted in the Emergency Department of a private-public Hospital in Naples, South of Italy (Buon Consiglio Fatebenefratelli). The Emergency Department serves the population which doesn't require specialistic care. The data collection has been performed in the period April–June 2019 before the Covid-19 pandemic. The research team has collected the data, observing directly the Triage assignment decision-making, made by different nurses, in different daily hours and in different days of the week. We have collected the priority codes assigned to 100 patients. These represent our cases in the application of fsQCA.

In order to apply the fsQCA we have, first of all, identified the conditions, that are the factors affecting the Triage process.

According to the analysis of literature [16, 31–39] and after an observation of the Triage, we selected the following factors:

- nurse experience measured as working years;
- the measurement of vital signs made by nurses to take a decision;
- number of nurses in each work shift;
- number of interruptions during the Triage process by colleagues;
- arrival of patients with ambulance;
- number of interruptions during the Triage process by patients.

These variables are related to individual, organizational and environmental factors and are the conditions of the fsQCA model.

The subsequent step concerned the definition of the outcome of the model. The outcome is related to the accuracy of the Triage process, affecting ED's quality. In fact, an under or over assessment of patients can increase the waiting time of urgent patients [40–44]. Over-assessment triage occurs if the patient receives a higher priority level than expected and this could affect the waiting time of more urgent patients. Under-assessment Triage implies that the patient has to wait more time to access to care than required. As output, we considered the under triage as it impacts directly on the waiting time of a specific patient while over triage can have an indirect impact on patients.

After defining conditions and the outcome, and after collecting data, we have performed the calibration phase using both fuzzy variables and crisp ones (Table 1). For crisp variables we have just the value "yes" or "not". Instead for fuzzy variables we have a membership function built on the characteristics of the Triage.

The calibration phase returns the following table of relationships between cases and conditions (Table 1. The variables used in QCA model. Table 2).

Based on the calibrated data, we proceed with constructing the "truth table", which proposes a list of all possible theoretical combinations of the causal conditions, the relative outcome, and the cases conforming to each combination [2]. The truth table treats each case as a combination of characteristics or 'configuration' in fsQCA terminology.

Then, the truth table is compared with the raw data table. The substantial difference between the two tables is the following: the first represents all possible configurations; the second, instead, report only the configurations emerged by empirical cases. In this way, we selected all the configurations of the truth table that we found in empirical cases. Furthermore, we simplified the truth table by applying a specific algorithm of fsQCA that minimizes the number of rows, eliminating those not relevant to explain the

**Table 1.** The variables used in QCA model.

| Variable | Acronym | Typology | Measurement | QCA Variable |
|---|---|---|---|---|
| Vital Signs | PV | Individual | Yes: if the decision is made measuring the vital signs No: otherwise | Crisp |
| Nurse' Experience | AE | Individual | Number of working years in health sector | Fuzzy |
| Number of nurses in each workshift | NT | Organizational | Yes: if 3 nurses are present in the workshift No: otherwise | Crisp |
| Number of interruptions during the Triage process by colleagues | NI | Organizational | Number of interruptions detected | Fuzzy |
| Arrival with ambulance | AM | Environmental | Yes: the patient arrives with ambulance No: otherwise | Crisp |
| Number of interruptions during the Triage process by patients | NP | Environmental | Number of interruptions detected | Fuzzy |
| Accuracy | OUTCOME | | Difference between priority code assigned by nurse and priority code reported on the patient's medical record | Fuzzy |

**Table 2.** Relationships between cases and conditions

| Em-pirical Cases (EC) | PV | AE | NP | AM | NI | NT | OUTCOME | Empiri-cal Cases (EC) | PV | AE | NP | AM | NI | NT | OUTCOME |
|---|---|---|---|---|---|---|---|---|---|---|---|---|---|---|---|
| EC 1 | 1 | 0.028 | 0.005 | 0 | 0.005 | 1 | 0 | EC 51 | 0 | 0.357 | 0.005 | 0 | 0 | 0 | 1 |
| EC 2 | 1 | 0.028 | 0.005 | 1 | 0 | 1 | 0.5 | EC 52 | 0 | 0.357 | 0.005 | 0 | 0.005 | 0 | 0 |
| EC 3 | 1 | 0.028 | 0.005 | 0 | 0 | 1 | 0 | EC 53 | 0 | 0.357 | 0.005 | 0 | 0.005 | 0 | 0 |
| EC 4 | 1 | 0.028 | 0.005 | 0 | 0 | 1 | 0.5 | EC 54 | 0 | 0.357 | 0.38 | 0 | 0.005 | 0 | 0 |
| EC 5 | 1 | 0.028 | 0.005 | 0 | 0 | 1 | 0.5 | EC 55 | 0 | 0.028 | 0.005 | 0 | 0 | 1 | 0 |
| EC 6 | 1 | 0.028 | 0.005 | 1 | 0 | 1 | 0 | EC 56 | 0 | 0.028 | 0.005 | 0 | 0 | 1 | 0 |
| EC 7 | 0 | 0.05 | 0.005 | 0 | 0 | 1 | 0 | EC 57 | 0 | 0.028 | 0.38 | 0 | 0 | 1 | 0 |
| EC 8 | 0 | 0.028 | 0.005 | 0 | 0.005 | 1 | 0 | EC 58 | 0 | 0.028 | 0.38 | 0 | 0.005 | 1 | 1 |
| EC 9 | 1 | 0.028 | 0.38 | 0 | 0.38 | 1 | 1 | EC 59 | 0 | 0.028 | 0.934 | 0 | 0.921 | 1 | 0 |
| EC 10 | 0 | 0.028 | 0.005 | 0 | 0 | 1 | 0 | EC 60 | 0 | 0.028 | 0.38 | 0 | 0.38 | 1 | 0 |
| EC 11 | 0 | 0.028 | 0.38 | 0 | 0 | 1 | 0 | EC 61 | 0 | 0.028 | 0.005 | 0 | 0.38 | 1 | 0 |
| EC 12 | 1 | 0.028 | 0.005 | 0 | 0 | 1 | 0.5 | EC 62 | 0 | 0.028 | 0.005 | 0 | 0 | 1 | 0 |
| EC 13 | 1 | 0.028 | 0.005 | 0 | 0 | 1 | 0 | EC 63 | 0 | 0.028 | 0.005 | 0 | 0.98 | 1 | 0 |
| EC 14 | 1 | 0.028 | 0.005 | 0 | 0.38 | 1 | 0.5 | EC 64 | 0 | 0 | 0.005 | 0 | 0 | 1 | 1 |
| EC 15 | 0 | 0.427 | 0.005 | 0 | 0 | 1 | 0 | EC 65 | 0 | 0 | 0.005 | 0 | 0.005 | 1 | 1 |
| EC 16 | 1 | 0.427 | 0.005 | 0 | 0 | 1 | 0.5 | EC 66 | 0 | 0 | 0.38 | 0 | 0 | 1 | 0 |
| EC 17 | 1 | 0.427 | 0.005 | 1 | 0.005 | 1 | 0 | EC 67 | 0 | 0.028 | 0.005 | 0 | 0.005 | 1 | 0 |
| EC 18 | 1 | 0.427 | 0.005 | 0 | 0 | 1 | 0 | EC 68 | 0 | 0.028 | 0.005 | 0 | 0 | 1 | 0 |
| EC 19 | 1 | 0.427 | 0.005 | 0 | 0.005 | 1 | 0.5 | EC 69 | 0 | 0.028 | 0.005 | 0 | 0.005 | 1 | 1 |
| EC 20 | 1 | 0.427 | 0.005 | 0 | 0.005 | 1 | 0 | EC 70 | 0 | 0.028 | 0.005 | 0 | 0.38 | 1 | 0 |
| EC 21 | 1 | 0.427 | 0.005 | 1 | 0.005 | 1 | 0 | EC 71 | 0 | 0.028 | 0.38 | 0 | 0.38 | 1 | 0 |
| EC 22 | 1 | 0.427 | 0.005 | 1 | 0.005 | 1 | 0 | EC 72 | 0 | 0.028 | 0.005 | 0 | 0.98 | 1 | 0 |
| EC 23 | 1 | 0.427 | 0.005 | 0 | 0.005 | 1 | 0 | EC 73 | 0 | 0.028 | 0.005 | 0 | 0 | 1 | 0 |
| EC 24 | 0 | 0.427 | 0.005 | 0 | 0 | 1 | 0.5 | EC 74 | 0 | 0.028 | 0.005 | 0 | 0.98 | 1 | 0 |
| EC 25 | 1 | 0.427 | 0.005 | 0 | 0.005 | 1 | 0.5 | EC 75 | 1 | 0.028 | 0.005 | 0 | 0.005 | 1 | 0.5 |

(*continued*)

**Table 2.** (*continued*)

| | | | | | | | | | | | | | | |
|---|---|---|---|---|---|---|---|---|---|---|---|---|---|---|
| EC 26 | 0 | 0.427 | 0.934 | 0 | 0 | 1 | 0.5 | EC 76 | 0 | 0.028 | 0.38 | 0 | 0.005 | 1 | 0 |
| EC 27 | 0 | 0.427 | 0.005 | 0 | 0 | 1 | 0 | EC 77 | 0 | 0.028 | 0.005 | 0 | 0 | 1 | 0.5 |
| EC 28 | 0 | 0.028 | 0.38 | 0 | 0 | 1 | 0 | EC 78 | 0 | 0.028 | 0.005 | 0 | 0 | 1 | 0 |
| EC 29 | 1 | 0.921 | 0.38 | 1 | 0.005 | 1 | 0 | EC 79 | 0 | 0.028 | 0.005 | 0 | 0.005 | 1 | 0 |
| EC 30 | 0 | 0.921 | 0.005 | 0 | 0 | 1 | 0 | EC 80 | 0 | 0.028 | 0.005 | 0 | 0.005 | 1 | 1 |
| EC 31 | 0 | 0.921 | 0.005 | 0 | 0.005 | 1 | 0 | EC 81 | 0 | 0.028 | 0.934 | 0 | 0 | 1 | 0 |
| EC 32 | 0 | 0.921 | 0.934 | 0 | 0.005 | 1 | 0.5 | EC 82 | 0 | 0.028 | 0.38 | 0 | 0.005 | 1 | 0 |
| EC 33 | 0 | 0.427 | 0.005 | 0 | 0.005 | 1 | 0 | EC 83 | 0 | 0.028 | 0.38 | 0 | 0.005 | 1 | 0 |
| EC 34 | 1 | 0.427 | 0.005 | 0 | 0 | 1 | 0.5 | EC 84 | 0 | 0.921 | 0.005 | 0 | 0 | 1 | 0.5 |
| EC 35 | 1 | 0.427 | 0.38 | 0 | 0 | 1 | 0.5 | EC 85 | 0 | 0.921 | 0.38 | 0 | 0.38 | 1 | 0 |
| EC 36 | 1 | 0.427 | 0.005 | 0 | 0 | 1 | 0.5 | EC 86 | 0 | 0.921 | 0.38 | 0 | 0.005 | 1 | 0 |
| EC 37 | 0 | 0.427 | 0.005 | 0 | 0.38 | 1 | 0 | EC 87 | 1 | 0.921 | 0.005 | 0 | 0 | 1 | 0 |
| EC 38 | 0 | 0.427 | 0.005 | 0 | 0.005 | 1 | 0 | EC 88 | 0 | 0.427 | 0.38 | 0 | 0.005 | 1 | 1 |
| EC 39 | 0 | 0.427 | 0.005 | 0 | 0 | 1 | 0 | EC 89 | 0 | 0.427 | 0.005 | 0 | 0 | 1 | 0.5 |
| EC 40 | 0 | 0.427 | 0.005 | 0 | 0 | 1 | 0.5 | EC 90 | 1 | 0.427 | 0.005 | 1 | 0 | 1 | 0 |
| EC 41 | 0 | 0.427 | 0.005 | 0 | 0 | 1 | 0 | EC 91 | 0 | 0.038 | 0.38 | 0 | 0 | 1 | 0 |
| EC 42 | 0 | 0.427 | 0.38 | 0 | 0.005 | 1 | 0 | EC 92 | 0 | 0.038 | 0.38 | 0 | 0 | 1 | 0.5 |
| EC 43 | 1 | 0.427 | 0.38 | 0 | 0 | 1 | 0.5 | EC 93 | 0 | 0.427 | 0.005 | 0 | 0.005 | 1 | 0 |
| EC 44 | 0 | 0.427 | 0.005 | 0 | 0 | 1 | 0 | EC 94 | 0 | 0.427 | 0.005 | 0 | 0 | 1 | 0 |
| EC 45 | 0 | 0.427 | 0.005 | 0 | 0 | 1 | 0 | EC 95 | 0 | 0.427 | 0.38 | 0 | 0.38 | 1 | 0 |
| EC 46 | 0 | 0.427 | 0.005 | 0 | 0 | 1 | 0 | EC 96 | 0 | 0.427 | 0.005 | 0 | 0.005 | 1 | 0 |
| EC 47 | 1 | 0.427 | 0.005 | 0 | 0 | 1 | 0.5 | EC 97 | 0 | 0.427 | 0.005 | 0 | 0 | 1 | 0 |
| EC 48 | 0 | 0.427 | 0.005 | 0 | 0.005 | 1 | 1 | EC 98 | 0 | 0.427 | 0.38 | 0 | 0.005 | 1 | 0.5 |
| EC 49 | 1 | 0.427 | 0.38 | 0 | 0.38 | 1 | 0 | EC 99 | 0 | 0.427 | 0.934 | 0 | 0.921 | 1 | 0 |
| EC 50 | 0 | 0.357 | 0.005 | 0 | 0.005 | 0 | 0 | EC 100 | 0 | 0.427 | 0.005 | 0 | 0 | 1 | 0 |

outcome. The minimization process returns the minimum number of configurations that explain the outcome (Fig. 1).

| | ~Outcome1 | | | | | |
|---|---|---|---|---|---|---|
| Conditions | 1 | 2 | 3 | 4 | 5 | 6 |
| PV | | ⊗ | | ⊗ | ● | ● |
| AE | ● | ⊗ | ⊗ | ⊗ | ● | |
| NI | ⊗ | | ⊗ | ● | ⊗ | ⊗ |
| NP | ⊗ | ● | ⊗ | | ⊗ | ⊗ |
| AM | ⊗ | ⊗ | ● | ⊗ | | ● |
| NT | ● | ● | ● | ● | ● | ● |
| Consistency | 0.934443 | 0.910119 | 0.890614 | 0.995636 | 0.996608 | 0.911528 |
| Raw coverage | 0.248679 | 0.0580879 | 0.0483396 | 0.0860881 | 0.103472 | 0.0641509 |
| Overall solution consistency | 0.939094 | | | | | |
| Overall solution coverage | 0.449371 | | | | | |
| Cases with greater than 0.5 membership in term | Caso30 (0.921,1), Caso31 (0.921,1), Caso84 (0.921,0.5), Caso87 (0.921,1), Caso85 (0.62,1), Caso86 (0.62,1) | Caso59 (0.934,1), Caso81 (0.934,1), Caso26 (0.573,0.5), Caso99 (0.573,1) | Caso2 (0.972,0.5), Caso6 (0.972,1), Caso17 (0.573,1), Caso21 (0.573,1), Caso22 (0.573,1), Caso90 (0.573,1) | Caso63 (0.972,1), Caso72 (0.972,1), Caso74 (0.972,1), Caso59 (0.921,1), Caso99 (0.573,1) | Caso87 (0.921,1), Caso29 (0.62,1) | Caso2 (0.995,0.5), Caso6 (0.995,1), Caso17 (0.995,1), Caso21 (0.995,1), Caso22 (0.995,1), Caso29 (0.62,1) |

**Fig. 1.** Solution for the outcome

In Fig. 1, the columns are the configuration of conditions emerged after the reduction of the truth table. For each configuration, we can identify if a condition is present or not in the specific term of the solution. The presence of a condition in a path is represented by the black circle, while its negation is a crossed circle. We have extracted 6 configurations. For the overall solution (made by six terms) and for each of them, we have the relative value of coverage and consistency in order to facilitate comparison. For the coverage, Ragin [25] does not propose any threshold value as it is linked to relevance empirical of the combination (how much of the outcome the combination can explain), whereas for the consistence the value must be around 0.8, but certainly not lower than 0.75. The results show that the combinations and the solution far exceed the value of the consistency threshold. In particular, the solution has an even consistency value at 94%. Therefore, the model has a strong explanatory power of the outcome. The value of coverage of the solution is instead equal to 45%, which indicates that 45% of the outcome "Is explained" by the solution: the solution was found in 45% of empirical cases.

## 4 Results

The results of the study show that the configuration presenting the highest coverage value (0.248679) and a good level of consistency (0.934) is the configuration (AE * ~ NP * ~ AM * ~ NI * NT). This configuration considers a triage process carried out by an expert nurse (AE), with the presence of all expected nurses in the Emergency Department area in the work shift analyzed (NT), in conditions of absence of interruptions by colleagues (~ NI) and by patients (~ NP), and patients not arrived by ambulance transportation (~ AM).

We can assume that the presence of all nurses and the absence of a crowd allow nurses to focus on specific patients, avoiding interruptions. Furthermore, an expert nurse can assign the correct priority code for patients not arriving with ambulance and without a pre-evaluation made by the ambulance's crew. In fact, all cases related to this configuration are registered by the same nurse, with 26 years of experience, and always in the morning, when the maximum number of nurses assigned to Triage is in the emergency room.

The configuration with the highest consistency level (0.996) is (PV * AE * ~ NP * ~ NI * NT), but it has a lower level of coverage than the previous combination (0.103). This configuration presents the measurement of vital signs (PV) by an expert nurse (AE), the absence of interruptions by patients (~ NP) and colleagues (~ NI), and the maximum number of nurses per shift (NT). This partially confirms the result described by the previous combination. In fact, this configuration also has the presence of the measurement of vital parameters. Although many studies state that more experienced nurses tend to do not measure vital signs [32, 34], in both cases related to this configuration, the detection of vital signs (PV) has been made. We can justify this choice by analyzing specific cases. In one case, the patient had abdominal pain, a symptom that requires the measurement of vital signs. In the other one, the arrival of the patient took place in an ambulance, so the vital signs had already been detected in the ambulance, but the long waiting time in the patient queue (12′30″) to access the Triage prompts the nurse to detect the vital signs again to grasp an evolution or changes in symptoms exhibited.

The other combinations confirm the presence of all nurses scheduled for the shift and the absence of interruptions by colleagues and patients. In a quiet environment, the nurse can detect the vital signs for the patients who arrived by ambulance to confirm the priority code assigned by the ambulance's crew or to capture any change in symptoms. In a quiet environment, also less expert nurses assign the correct priority code to the patient.

According to these 4 out of 6 configurations, we can infer that the Triage should be performed in a dedicated room avoiding interruptions and that the measurement of vital signs could be a valuable support for the accuracy even for expert nurses that usually make decisions based primarily on their intuition.

The other two configurations with a lower value of coverage and consistency confirm that the presence of three nurses in the Emergency Department is an ever-present condition; unlike the other four configurations, these last two presents the lack of experience (~ AE), the scarce use of vital parameters (~ PV), and the evaluation of patients who arrived at the ED independently (~ AM). Such configurations suggest that, despite the lack of experience, the nurse relies mainly on their "intuition" rather than on vital parameters in evaluating patients arriving independently. Additionally, it seems that the presence of all scheduled nurses in a work shift can instill a greater sense of security in less expert nurses who can always rely on more experienced colleagues. However, less experienced nurses take more time to assess the patients. In fact, a more profound analysis of the cases relative to these two configurations showed a high number of questions for the attribution of the code, confirming what emerged by the literature: less experienced nurses collect more information from the patient to formulate the Triage decision [45].

These results suggest some managerial interventions on the Triage process. Firstly, the accuracy of Triage can be improved if almost an experienced nurse is devoted exclusively to the Triage. In this way, he/she can concentrate fully on the Triage process without external pressure. Furthermore, the great experience allows the nurse to make an accurate decision even for walking patients, which are not pre-assessed for their clinical conditions. This suggests a redesign of the layout of the analyzed Hospital, with a room dedicated exclusively to Triage with an expert nurse in crowded situations or less expert nurses in quiet situations since they take longer to assign the priority code. Concerning the organizations of the work shifts, our results show that the number of Triage nurses simultaneously present in the Emergency Department has to be the maximum number available and that disparities among the work shifts have to be removed.

Another interesting result concerns the vital signs that are not the main information used by nurses. These findings could encourage the Emergency Department to adopt tele-Triage or pre-Triage to better address the patient towards Emergency Departments or other structures to face the social distancing due to COVID-19.

## 5 Conclusions

This paper presents the application of fuzzy Qualitative Comparative Analysis to the dynamic decision-making in Triage assessment. Unlike other works, this research is carried out through direct observation of the Triage process in its empirical reality; this allowed to consider also environmental factors in the study.

Findings suggest that the accuracy of the Triage process requires that an experienced nurse carries out the assignment process in a situation in which all scheduled nurses are present in the work shift with the absence of interruptions by colleagues and by patients. All configurations confirm as relevant the presence of all nurses in the work shifts, whereas the measurement of vital signs can be helpful support for the accuracy even if the nurses rely primarily on their intuition.

Following a methodological perspective, this application shows that fsQCA is an effective method for understanding the cognitive heuristic decision-making, highlighting how individual, organizational and environmental factors interact for producing an effective Triage.

From a practical perspective, these findings suggest that the accuracy of Triage can be improved if an experience nurse is devoted exclusively to the Triage. In this way he/she can concentrate fully on the Triage process, without external pressure. Furthermore, the great experience allows the nurse to make an accurate decision even for the walking patients, which are not a pre-assessed for their clinical conditions. Another interesting result concerns the vital signs that are not the main information used by nurses. These findings could encourage the Emergency Department in adopting solution of tele-Triage or pre-Triage to better address the patient towards Emergency Departments or other structures, in order to face the social distancing due to COVID-19. The study has been already used to organize a training session for the nurses of the Hospital involved in the research. Unfortunately, the advent of the outbreak has not allowed to complete the process of training and also the discussion with managers of the Emergency Department to design organizational interventions on Triage.

## References


1. Kraus, S., Ribeiro-Soriano, D., Schüssler, M.: Fuzzy-set qualitative comparative analysis (fsQCA) in entrepreneurship and innovation research–the rise of a method. Int. Entrep. Manag. J. **14**(1), 15–33 (2018)
2. Ragin, C.: Fuzzy-set social science. University of Chicago Press, Chicago, IL (2000)
3. Rihoux, B.: Qualitative comparative analysis (QCA) and related systematic comparative methods: Recent advances and remaining challenges for social science research. Int. Sociol. **21**(5), 679–706 (2006)
4. Marx, A., Cambré, B., Rihoux, B.: Chapter 2 crisp-set qualitative comparative analysis in organizational studies. In: Fiss, P.C., Cambré, B., Marx, A. (eds.) Configurational theory and methods in organizational research, pp. 23–47. Emerald Group Publishing Limited (2013). https://doi.org/10.1108/S0733-558X(2013)0000038006
5. Misangyi, V.F., Greckhamer, T., Furnari, S., Fiss, P.C., Crilly, D., Aguilera, R.: Embracing causal complexity: the emergence of a neo-configurational perspective. J. Manag. **43**(1), 255–282 (2017)
6. Woodside, A.G., Nagy, G., Megehee, C.M.: Applying complexity theory: A primer for identifying and modeling firm anomalies. J. Innov. Knowl. **3**(1), 9–25 (2018)
7. Simon, H.A., Newell, A.: Human problem solving: the state of the theory in 1970. Am. Psychol. **26**(2), 145 (1971)
8. Basel, J.S., Brühl, R.: Rationality and dual process models of reasoning in managerial cognition and decision making. Eur. Manag. J. **31**(6), 745–754 (2013)
9. Weeks, D., Whimster, S.: Contexted decision making. In: Behavioral Decision Making, pp. 167–188. Springer, Boston, MA (1985)
10. Haley, U.C., Stumpf, S.A.: Cognitive trails in strategic decision-making: linking theories of personalities and cognitions. J. Manage. Stud. **26**(5), 477–497 (1989)
11. Wilson, T.: Exploring models of information behaviour: the 'uncertainty' project. Inf. Process. Manage. **35**(6), 839–849 (1999)
12. Pachur, T., Galesic, M.: Strategy selection in risky choice: the impact of numeracy, affect, and cross-cultural differences. J. Behav. Decis. Mak. **26**(3), 260–271 (2013)
13. Noon, A.J.: The cognitive processes underpinning clinical decision in triage assessment: a theoretical conundrum? Int. Emerg. Nurs. **22**(1), 40–46 (2014)
14. Carnevali, D.L., Mitchell, P.H., Woods, N.F., Christine, A.: Diagnostic reasoning in nursing. AJN The Am. J. Nurs. **85**(7), 838 (1985)
15. Benner, P., Tanner, C.: How expert nurses use intuition. AJN The Am. J. Nurs. **87**(1), 23–34 (1987)
16. Andersson, A.K., Omberg, M., Svedlund, M.: Triage in the emergency department-a qualitative study of the factors which nurses consider when making decisions. Nurs. Crit. Care **11**(3), 136–145 (2006)
17. Göransson, K.E., Ehnfors, M., Fonteyn, M.E., Ehrenberg, A.: Thinking strategies used by Registered Nurses during emergency department triage. J. Adv. Nurs. **61**(2), 163–172 (2008)
18. Cone, K.J., Murray, R.: Characteristics, insights, decision making, and preparation of ED triage nurses. J. Emerg. Nurs. **28**(5), 401–406 (2002)
19. Terris, J., Leman, P., O'connor, N., Wood, R.: Making an IMPACT on emergency department flow: improving patient processing assisted by consultant at triage. Emerg. Med. J. **21**(5), 537–541 (2004)
20. Rodi, S.W., Grau, M.V., Orsini, C.M.: Evaluation of a fast track unit: alignment of resources and demand results in improved satisfaction and decreased length of stay for emergency department patients. Qual. Manag. Healthcare **15**(3), 163–170 (2006)

21. Ashour, O.M., Okudan Kremer, G.E.: Dynamic patient grouping and prioritization: a new approach to emergency department flow improvement. Health Care Manag. Sci. **19**(2), 192–205 (2014)
22. Elalouf, A., Wachtel, G.: An alternative scheduling approach for improving patient-flow in emergency departments. Oper. Res. Health Care **7**, 94–102 (2015)
23. Wouters, L.T., et al.: Tinkering and overruling the computer decision support system: working strategies of telephone triage nurses who assess the urgency of callers suspected of having an acute cardiac event. J. Clin. Nurs. **29**(7–8), 1175–1186 (2020)
24. Ragin, C.C.: The comparative method. In: Moving Beyond Qualitative and Quantitative Strategies. University of California Press, Berkeley, CA (1987)
25. Ragin, C.C.: Redesigning Social Inquiry: Fuzzy Sets and Beyond, vol. 240. University of Chicago Press, Chicago, IL (2008)
26. Schneider, C.Q., Wagemann, C.: Qualitative comparative analysis (QCA) and fuzzy-sets: agenda for a research approach and a data analysis technique. Comp. Sociol. **9**(3), 376–396 (2010)
27. Legewie, N.: An introduction to applied data analysis with qualitative comparative analysis. In: Forum Qualitative Sozialforschung/Forum: Qualitative Social Research, vol. 14, issue 3, Art. 15 (2013)
28. Ragin, C.C.: The limitations of net-effects thinking. In: Rihoux, B., Grimm, H. (eds.) Innovative Comparative Methods for Policy Analysis, pp. 13–41. Springer, New-York (2006)
29. Ragin, C.C.: Qualitative comparative analysis using fuzzy sets (fsQCA). In: Rihoux, B., Ragin, C.C. (eds.) Configurational Comparative Methods: Qualitative Comparative Analysis (QCA) and Related Techniques, pp. 87–121. SAGE Publications, Thousand Oaks, CA (2009)
30. Chang, M.L., Cheng, C.F.: How balance theory explains high-tech professionals' solutions of enhancing job satisfaction. J. Bus. Res. **67**(9), 2008–2018 (2014)
31. Considine, J., Botti, M., Thomas, S.: Do knowledge and experience have specific roles in triage decision-making? Acad. Emerg. Med. **14**(8), 722–726 (2007). https://doi.org/10.1197/j.aem.2007.04.015
32. Chung, J.Y.M.: An exploration of accident and emergency nurse experiences of triage decision making in Hong Kong. Accid. Emerg. Nurs. **13**(4), 206–213 (2005)
33. Cooper, R.J., Schriger, D.L., Flaherty, H.L., Lin, E.J., Hubbell, K.A.: Effect of vital signs on triage decisions. Ann. Emerg. Med. **39**(3), 223–232 (2002). https://doi.org/10.1067/mem.2002.121524
34. Gerdtz, M.F., Bucknall, T.K.: Triage nurses' clinical decision making. An observational study of urgency assessment. J. Adv. Nurs. **35**(4), 550–561 (2001). https://doi.org/10.1046/j.1365-2648.2001.01871.x
35. Ponsiglione, C., Ippolito, A., Primario, S., Zollo, G.: Configurations of factors affecting triage decision-making: a fuzzy-set qualitative comparative analysis. Manag. Decis. **56**(10), 2148–2171 (2018). https://doi.org/10.1108/MD-10-2017-0999
36. Smith, M., Higgs, J., Ellis, E.: Factors influencing clinical decision making. In: Higgs, J., et al. (eds.) Clinical Reasoning in the Health Professions, pp. 89–100. Elsevier, Edinburgh (2008)
37. Soremekun, O.A., Takayesu, J.K., Bohan, S.J.: Framework for analyzing wait times and other factors that impact patient satisfaction in the emergency department. J. Emerg. Med. **41**(6), 686–692 (2011). https://doi.org/10.1016/j.jemermed.2011.01.018
38. van der Linden, M.C., Meester, B.E.A.M., van der Linden, N.: Emergency department crowding affects triage processes. Int. Emerg. Nurs. **29**, 27–31 (2016). https://doi.org/10.1016/j.ienj.2016.02.003
39. Johnson, K.D., Motavalli, M., Gray, D., Kuehn, C.: Causes and Occurrences of Interruptions During ED Triage. J. Emerg. Nurs. **40**(5), 434–439 (2014). https://doi.org/10.1016/j.jen.2013.06.019

40. Fernandes, C.M., et al.: Five-level triage: a report from the ACEP/ENA five-level triage task force. J. Emerg. Nurs. **31**(1), 39–50 (2005)
41. Considine, J., LeVasseur, S.A., Villanueva, E.: The Australasian Triage Scale: examining emergency department nurses' performance using computer and paper scenarios. Ann. Emerg. Med. **44**(5), 516–523 (2004)
42. Wuerz, R., Fernandes, C.M.B., Alarcon, J.: Inconsistency of emergency department triage. Ann. Emerg. Med. **32**(4), 431–435 (1998). https://doi.org/10.1016/S0196-0644(98)70171-4
43. Fitzgerald, M., Gibson, F., Gunn, K.: Contemporary issues relating to assessment of pre-registration nursing students in practice. Nurse Educ. Pract. **10**(3), 158–163 (2010)
44. Traub, S.J., et al.: Emergency department rapid medical assessment: overall effect and mechanistic considerations. J. Emerg. Med. **48**(5), 620–627 (2015)
45. Cioffi, J.: Decision making by emergency nurses in triage assessments. Accid. Emerg. Nurs. **6**(4), 184–191 (1998)

# ESG Risk Disclosure and Earning Timelines in the Mexican Capital Market Using Fuzzy Logic Regression

Martha del Pilar Rodríguez García(✉)

Facultad de Contaduría Pública y Administración, Universidad Autónoma de Nuevo León, 66451 San Nicolás de los Garza, NL, Mexico
martha.rodriguezgc@uanl.edu.mx

**Abstract.** This study analyzes whether ESG risk information disclosures impact on earnings per share to confirm accounting conservatism in Mexico. The main objective is to determine if nonfinancial information disclosure, such as the ESG risk metric, affects the quality the earning timeliness in Mexico. This re-search was conducted during 2020 using a sample of 31 listed companies from the Mexican capital market. However, due to the volatility and uncertainty of the 2020 profits due to the COVID-19 pandemic crisis, an alternative method to estimate under this environment is necessary. Thus, to estimate the model, we applied the fuzzy regression. The results suggest that the impact of the market reaction (positive or negative returns) on the financial performance is ambiguous since there is a probability of 41% to obtain a positive or direct effect of market returns on financial performance. However, this probability rises to 100% if the firms increase the ESG risk to medium, high, or severe levels. On other words, the "bad socially responsible news" increases the impact of the market reaction to financial performance. These results suggest the presence of accounting conservatism in the Mexican capital market in 2020 due to the negative disclosures of ESG risks.

**Keywords:** Accounting conservatism · Corporate social responsibility · Quality financial information

## 1 Introduction

As Corporate Social Responsibility (CSR) has become more important in our society, many companies have decided to use social and ethics manuals to disclose their social activities to project a positive corporate image to the society and industry in which they function. Currently, CSR is accompanied by many external and internal factors, including the attitude and participation of company employees and the corporate social programs instituted by the government that promote the creation of new, responsible businesses.

The accounting information has been used to understand the economic performance of the firm to make decisions [1]. The modern economy has been characterized by corporate social responsibility (CSR) practices, as well as pursuing corporate profit, and


M. d. P. Rodríguez García et al. (Eds.): XX SIGEF 2021, LNNS 384, pp. 73–81, 2022.
https://doi.org/10.1007/978-3-030-94485-8_6

reflecting the actual corporate disclosure trends [2]. High standards of CSR disclosure can help corporate managers establish a positive image of the firm, reduce its market risk, and increase their stock price [3]. Similarly, voluntary firm disclosure of nonfinancial information also has corporate benefits by minimizing information asymmetry and maximizing the allocation of resources for investors [4]. This scenario enables investors to evaluate firms' potential risk and reduce the cost of [5].

The conservatism analysis has been used to analyze sustainability management. For instance, Hong [2] demonstrates that higher ESG rating in listed South Korean firms induce a lower conservatism. As a difference of previous studies, our research analyses the relationship between ESG risk metrics and accounting conservatism. Basu [6] commented that, since accountants anticipate future losses but not future profits, conservatism results in earnings being more timely and more sensitive to publicly available 'bad news' than 'good news'. In this sense, if we considered ESG risk measure, the earning per share is predicted to be more strongly associate with negative measure of ESG risk as a "proxy of bad news" than positive returns which is a "proxy of good news".

This research is motived by empirical evidence of the ESG disclosure impact on earnings per share to measure the accounting conservatism in Mexico. To determine the presence of accounting conservative we proposed a model to assess whether the information disclosure of ESG risk is relevant for market investors, that is, the presence of bad news of ESG have a greater impact on earing per shares to price ratio than the absence of them. However, due to the volatility and uncertainty of the 2020 profits due to the COVID-19 pandemic crisis, an alternative method to estimate under this environment is necessary. Thus, to estimate the model, we applied the fuzzy regression method proposed by Sakawa and Yano [7].

The main objective of this research is to determine if nonfinancial information disclosure, such as the ESG risk metric, affects the quality the earning timeliness in Mexico during 2020. The findings of this study suggest the presence of accounting conservatism measure by the timeliness of earnings because ESG risk disclosure impact in the earning per share in Mexican firms. Thus, there is evidence that the quality of accounting information increases with ESG risk metrics disclosure adoption.

The findings of this study suggest the presence of accounting conservatism measure by the timeliness of earnings because ESG risk disclosure impact in the earning per share in Mexican firms. Thus, there is evidence that the quality of accounting information increases with ESG risk metrics disclosure adoption.

This paper is organized into five sections. Section 2 presents the theoretical framework for eco-efficiency and a review of the studies on the relationship be-tween financial performance and eco-efficiency. In Sect. 3, we present the data and methods. Finally, Sects. 4 and 5 present the study's results and conclusions, respectively.

## 2 Theorical Background

Accounting conservatism could be explained by the rule "anticipate no profits but anticipate all losses" [8]. Basu [6] interpret conservatism as capturing accountants' tendency to require a higher degree of verification for recognizing good news than bad news in financial statements, that is, earnings reflect bad news more quickly than good news.

Shen et al. [9] mentioned that accounting conservatism effectively improves the information disclosure environment and investor protection, this is because reduces information asymmetry and agency costs.

Barth et al. [10] state that financial information quality is presented by three reasons: the accounting information is less manipulated, a more opportune acknowledgment of the losses is presented, and an increase in the predictive capability is given by the regression between the fundamental and market variables. Other authors, such as Ball et al. [11] and Burgstahler et al. [12] mention that the accounting information quality varies when the number of users have access to privileged information. Despite the lack of a clear definition of the accounting quality, several studies use measures that are considered substitutes of accounting quality, for instance, the earning management, the opportune acknowledgment of losses (timeliness earning) and the evaluative relevance [10].

In this regard, Brennan and Tamarowski [13] suggested that transparent dis-closures and financial statements reduce the cost of information acquisition, and it has a positive effect on stock liquidity. Moreover, the disclosure of mandatory CSR reports increases the number of analysts following a company and diminishes market information asymmetry [14, 15].

Hence, the ESG disclosure could be a complement of financial accounting and should increase financial performance [16]. Velte [17] mention that financial and non-financial information disclosure will be included in an overall stakeholder communication strategy. The degree of financial reporting quality and ESG dis-closure quality as key management decisions will be connected.

The companies that report CSR performance have a stronger motivation than previously to improve the quality of their financial statements because of ethical issues [18, 19]. Modern accounting standards emphasize non-financial indicators as a means of compensating for weaknesses in financial indicators. Non-financial information supplements financial information and can also affect financial information disclosure [2].

# 3 Methodology

## 3.1 Data and Sample

The present research was conducted in 2020 using a sample of 31 listed companies in the Mexican stock exchange market. The firms of the sample operate in financial and non-financial sectors. For this study, we used the Koyfin database (https://www.koyfin.com/) to obtain the quarterly series of the accounting variables and markets prices.

On the other hand, the ESG risk metrics of the selected firms were obtained by the web platform of Sustainalitycs (https://www.sustainalytics.com/). The ESG risk is measured in five levels (see Table 1):

1. Negligible risk, enterprise value is considered to have a negligible risk of material financial impacts driven by ESG factors.
2. Low risk, enterprise value is considered to have a low risk of material financial impacts driven by ESG factors.
3. Medium risk, enterprise value is considered to have a medium risk of material financial impacts driven by ESG factors.

4. High risk, enterprise value is considered to have a high risk of material financial impacts driven by ESG factors.
5. Severe risk, enterprise value is considered to have a severe risk of material financial impacts driven by ESG factors.

**Table 1.** Sample and ESG risk.

| Market ticker | Firms with medium, high, or severe risk | Market ticker | Firms with low, negligible, or not reported risk |
|---|---|---|---|
| ALFAA | Alfa | ALSEA | Alsea |
| AMXL | America Movil | AC | Arca Continental |
| KOFUBL | Coca-Cola Femsa | BBAJIOO | Banco del Bajio |
| FEMSAUBD | Fomento Economico Mexicano | CUERVO | Becle |
| GRUMAB | Gruma | BOLSAA | Bolsa Mexicana de Valores |
| GAPB | Grupo Aeroportuario del Pacifico | VESTA | Corp Inmobiliaria Vesta |
| BIMBOA | Grupo Bimbo | LIVEPOLC | El Puerto de Liverpool |
| GCARSOA1 | Grupo Carso | LABB | Genomma Lab Internacional |
| GFNORTEO | Grupo Financiero Banorte | OMAB | Grupo Aeroportuario del Centro |
| GFINBURO | Grupo Financiero Inbursa | ASURB | Grupo Aeroportuario del Sureste |
| GMEXICOB | Grupo Mexico | GCC | Grupo Cementos de Chihuahua |
| PE&OLES | Industrias Penoles | ELEKTRA | Grupo Elektra |
| IENOVA | Infraestructura Energetica Nova | PINFRA | Promotora y Operadora de Infraestructura |
| KIMBERA | Kimberly-Clark de Mexico | Q | Qualitas Controladora |
| ORBIA | Orbia Advance Corp | SITESB1 | Telesites |
| ALFAA | Alfa | WALMEX | Wal-Mart de Mexico |
| | | ALSEA | |

## 3.2 Empirical Model

To measures the asymmetrical timeliness of earnings we used the model suggested by Basu [6] which established that the earnings capture the "bad news" quicker than the "good news" due to the existence of an asymmetry in the verification standards of the losses and profits. In this idea we expected that if we considered ESG risk measure, the earning per share is predicted to be more strongly associate with negative measures of ESG risk or "proxy of bad news" than positive returns which are more identify for "good news".

To verify the relationship between CRS and accounting conservatism, research models are designed, then following [2] and [6], the basic model of the effects of expected returns on profits is:

$$X_i/P_i = \alpha + \beta R_i + e_i \tag{1}$$

where,

$X_i =$ are the earnings per share for the *i-th* firm of the year 2020.

$P_i =$ is the price per share for the *i-th* firm at the beginning of the year 2020.

$R_i =$ is the stock return for the *i-th* firm cumulated over the year 2020.

Equation (1) considers that the stock returns impact the sensitivity performance on profits. In other words, how fast earnings react to good and bad news of the market. Therefore $\beta$ should be significant and different from zero. However, to test the presence of earnings timeliness with ESG risk disclosure using the model proposed by Hong [2], i.e., if bad news is more sensitive than good news then the slope of the Eq. (1) can be disaggregated as follow:

$$\beta = \beta_1 + \beta_2 D_i \tag{2}$$

where,

$D_i =$ Dichotomous variable that takes the value of 1 if ESG risk metrics is medium, high, and severe risk, and 0 otherwise, that is, low and negligible ESG risk.

The model used to test the reaction of the profits to bad news and the incorporation of ESG risk information is obtaining substituting Eq. (2) in Eq. (1) as it is indicated below:

$$X_i/P_i = \alpha + \beta_1 R_i + \beta_2 D_i R_i + e_i \tag{3}$$

As a result, if there is evidence that the ESG risk disclosure increase the earnings timeliness in México during 2020 then $\beta_2$ must be positive, and bigger than $\beta_1$. To estimate $\beta_1$ and $\beta_2$ we considered the fuzzy regression methods proposed by Sakawa and Yano [7].

### 3.3 Fuzzy Estimation Regression

In the literature, the fuzzy regression has been applied to finance issues [20, 21]. Fuzzy regression models seek to determine functional relationships between a dependent variable $Y$ (also called the response variable, for instance, earning per share to price ratio) and one or more independent (explanatory as returns and ESG risk) variables, $X = (X_0, X_1, \ldots, X_m)$, for which $X_0 = 1$ and $m$ is the number of parameters to estimate using confidence intervals (CIs). A CI is represented either by its minimum ($a_1$) and maximum ($a_2$) values as $A = [a_1, a_2]$ or through its center ($a_c$) and radius ($a_R$) as follows:

$$a_C = \frac{a_2 + a_1}{2} \qquad a_R = \frac{a_2 - a_1}{2} \tag{4}$$

where the $m$ parameters of the regression are CIs of the form $A_i = \langle a_{iC}, a_{iR} \rangle$ with $i = 1, 2, \ldots, m$.

For a particular phenomenon, assumed that the observer has a sample represented as $(Y_1, X_1), \ldots, (Y_n, X_n)$, Sakawa and Yano [7] believed that the observations with $j = 1, \ldots, n$ must be "equal" to their estimates, i.e., $Y_j = \hat{Y}_j$ follows the next conditions:

$$Y_{jC} + Y_{jR} \geq \hat{Y}_{jC} - \hat{Y}_{jR} \text{ and } Y_{jC} - Y_{jR} \leq \hat{Y}_{jC} + \hat{Y}_{jR} \tag{5}$$

where the parameters are estimated by solving the following linear programming problem:

$$Min\ z = \sum_{j=1}^{n} \hat{Y}_{jR} = \sum_{j=1}^{n} \sum_{i=0}^{m} a_{iR} |X_{ij}| \tag{6}$$

subject to:

$$\hat{Y}_{jC} - \hat{Y}_{jR} = \sum_{i=0}^{m} a_{iC} X_{ij} - \sum_{i=0}^{m} a_{iR} |X_{ij}| \leq Y_{jC} + Y_{jR}$$

$$\hat{Y}_{jC} + \hat{Y}_{jR} = \sum_{i=0}^{m} a_{iC} X_{ij} + \sum_{i=0}^{m} a_{iR} |X_{ij}| \geq Y_{jC} - Y_{jR}$$

$$a_{iR} \geq 0; i = 0, 1, \ldots, m; j = 1, 2, \ldots, n$$

where,

z = is the total uncertainty within the sample measured as the sum of the radii of the estimates.

## 4 Results

Figure 1 shows the fuzzy estimation regression in contrast to real information. The estimation fuzzy financial performance covers the observed data in all the sample. Special cases are notary on GMEXICO and PE&OLES, both firms dedicated mainly in the mineral sector with bigger uncertainty.

The results suggest that the impact of the market reaction (positive or negative returns) on the financial performance is ambiguous since de fuzzy beta ($\beta_1$) estimated takes values between [–6.56%; 4.46%], see Table 2. In this regard, there is a probability of 41%, i.e., 4.46/(6.56 + 4.46), to obtain a positive or direct effect of market returns on financial performance. However, this probability increases to 100% if the firms increase the ESG risk to medium, high, or severe levels. On other words, the "bad socially responsible news" increases the impact of the market reaction to financial performance since the fuzzy beta of ESG risk market sensibility, $\beta_2 =$ [6.66%; 42.79], is positive and higher than $\beta_1$. These results suggest the presence of accounting conservatism in the Mexican capital market in 2020 due to the negative disclosures of ESG risks.

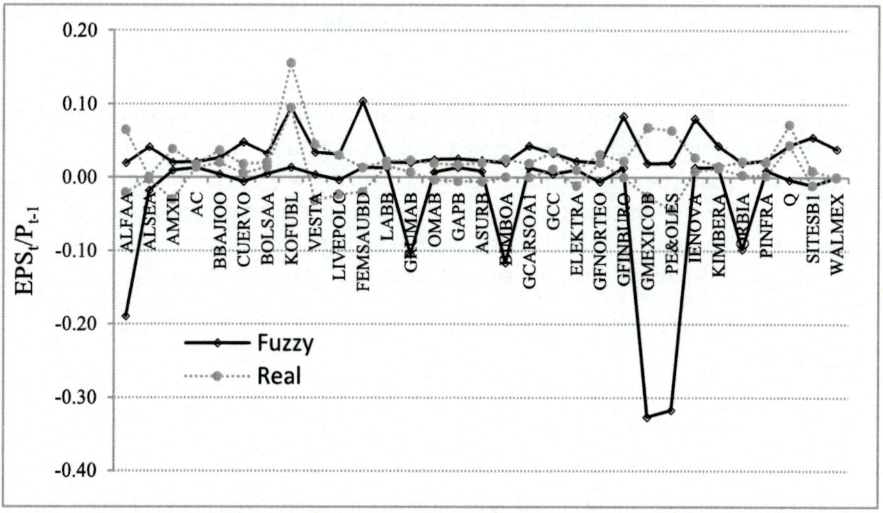


**Fig. 1.** Fuzzy financial performance estimations versus observed real data.

**Table 2.** Fuzzy regression estimations.

| Beta estimation | Fuzzy beta (central and radii) | | Fuzzy beta (range) | |
|---|---|---|---|---|
| | $\beta^c$ | $\beta^R$ | $\beta^{min}$ | $\beta^{max}$ |
| $\beta_0$ | 1.65% | 0.34% | 1.31% | 1.99% |
| $\beta_1$ | −1.05% | 5.51% | −6.56% | 4.46% |
| $\beta_2$ | 24.73% | 18.07% | 6.66% | 42.79% |

## 5 Conclusions

This study is motived by study the effect of ESG risk in the accounting measures to probe is, the accounting conservatism presented in Mexican firms in uncertainty times during 2020. The contributions of this study are relevant for at least three reasons. First, due to the lack of literature on ESG risk and quality accounting measures as such as earning timeliness [2]. Second, analysis is presented in Mexico, and thus, the quality of information is analyzed considering ESG risk metrics and the market performance of the Mexicans firms. Third, the effect of ESG risk on the quality of accounting information is tested considering earning timeliness using fuzzy regression.

H1 shows that in the total sample, there is asymmetrical timeliness of earnings. On the other hand, we found that ESG risk disclosure shows asymmetrical earnings timeliness in Mexico during 2020. This conclusion is like that of [9] who find that the ESG disclosure. To support the hypothesis of earnings timeliness, we estimated a medicated Ohlson model, and our results show earnings timeliness was presented in Mexico.

Financial statements and nonfinancial information in Mexico that consider legal, fiscal, social, and cultural differences, as well as different criteria for recognition, valuation, and publication, directly affect the comprehension and comparability of financial information at the international level. In Mexico companies with higher ESG risk have higher accounting conservatism measure with earning timeliness. Hong [2] states that a high-quality information disclosure is more inclined to adopt conservative accounting policies to reduce information asymmetry between internal managers and externa stakeholders. Our research reveals that companies with high ESG risk scores could affect the earning price ratio when this ESG risk are medium, high, and severe risk.

## References

1. Dumitru, M., Calu, D.A., Gorgan, C., Calu, A.: A historical approach of change in management accounting topics published in Romania. Account. Manag. Inf. Syst. **10**(3), 375–396 (2011)
2. Hong, S.: Corporate social responsibility and accounting conservatism. Int. J. Econ. Bus. Res. **19**(1), 1–18 (2020)
3. Kim, B.Y., Park, H.J.: The impact of accounting conservatism and foreign ownership on agency costs. Korean Account. J. **23**(5), 145–175 (2014)
4. Healy, P.M., Palepu, K.G.: Information asymmetry, corporate disclosure, and the capital markets: a review of the empirical disclosure literature. J. Account. Econ. **31**, 405–440 (2001)
5. Albarrak, M.S., Elnahass, M., Salama, A.: The effect of carbon dissemination on cost of equity. Bus. Strateg. Environ. **28**, 1179–1198 (2019)
6. Basu, S.: The conservatism principle and the asymmetric timeliness of earnings. J. Account. Econ. **24**(3), 3–37 (1997)
7. Sakawa, M., Yano, H.: Fuzzy regression and its applications. In: Kacprzyk, J., Fedrizzi, M. (eds.) Fuzzy regression analysis, pp. 91–101. Physica-Verlag, Heidelberg (1992)
8. Bliss, J.H.: Management through Accounts. The Ronald Press Company, New York (1924)
9. Shen, X., Ho, K.C., Yang, L., Wang, L.F.S.: Corporate social responsibility, market reaction and accounting conservatism. Kybernetes **50**(6), 1837–1872 (2021)
10. Barth, M.E., Landsman, W.R., Lang, M.H.: International accounting standards and accounting quality. J. Account. Res. **46**, 467–498 (2008)
11. Ball, R., Robin, A., Wu, J.S.: Incentives versus standards: properties of accounting income in four EastAsian countries, and implications for acceptance of IAS. J. Account. Econ. **36**(1–3), 235–270 (2003)
12. Burgstahler, D., Hail, L., Leuz, C.: The importance of reporting incentives: earnings management in Europeanprivate and public firms. Account. Rev. **81**(5), 983–1016 (2006)
13. Brennan, M.J., Tamarowski, C.: Investor relations, liquidity, and stock prices. J. Appl. Corp. Financ. **12**(4), 26–37 (2000)
14. Hung, M., Shi, J., Wang, Y.: The effect of mandatory CSR disclosure on information asymmetry: evidence from a quasi-natural experiment in China. In: Asian Finance Association (AsFA) Conference (2013)
15. Callen, J.L., Khan, M., Lu, H.: Accounting quality, stock price delay, and future stock returns. Contemp. Account. Res. **30**(1), 269–295 (2013)
16. Murphy, D., McGrath, D.: ESG reporting – class actions, deterrence, and avoidance. Sustain. Account. Manag. Policy J. **4**(2), 216–235 (2013)
17. Velte, P.: The bidirectional relationship between ESG performance and earnings management–empirical evidence from Germany. J. Glob. Responsib. **10**(4), 322–338 (2019)

18. Kim, Y., Park, M.S., Wier, B.: Is earnings quality associated with corporate social responsibility? Account. Rev. **87**(3), 761–796 (2012)
19. Bereskin, F., Byun, S.K., Officer, M.S., Oh, J.M.: The effect of cultural similarity on mergers and acquisitions: evidence from corporate social responsibility. J. Financ. Quant. Anal. **53**(5), 1995–2039 (2018)
20. Terceño, A., Barberà, G., Vigier, H., Laumann, Y.: Coeficiente Beta en sectores del mercado español. Regresión borrosa vs regresión ordinaria. Cuadernos del CIMBAGE **13**, 79–105 (2011)
21. De Los Cobos-Silva, S., Goddard-Close, J., Gutierrez-Andrade, M.: Regresión borrosa vs. regresión por mínimos cuadrados ordinarios: caso de estudio. Revista de Matemática Teoría y Aplicaciones **18**(1), 33–48 (2011)

# The Digital Taxation Adoption and Its Impact on Income Tax in Mexico (2010–2020)

Fabiola Denisse Flores-Guajardo(✉), Juan Paura-García, and Daniel Oswaldo Flores-Silva

Facultad de Contaduría Pública y Administración, Universidad Autónoma de Nuevo León, Av. Universidad S/N, Cd. Universitaria, 66451 San Nicolás de los Garza, NL, Mexico
fdenisse.flores@gmail.com

**Abstract.** This research analyzes the Income Tax Collection in Mexico derived from the evolution of the Digital Taxation mechanisms (MHCR) that includes the electronic bill, the electronic signature, the tax mailbox, the electronic accounting and tax payments referenced, whose first sign of the new digital era was in 2000, but its use became more acute as of the 2014 Mexican Tax Re-form. The new adaptations and way of analyzing the information have been made in order to increase the country's tax collection and in turn generate greater control of taxpayer's operations, so these mechanisms allow them to comply with their tax obligations easier and more confidence. In this research, the Income Tax collected in Mexico during the years 2010 to 2020 is analyzed by means of a factor analysis to determine a variable that represents the economic activity of the country and a multiple linear regression for the adapted Cobb-Douglas model with data from the Mexican Ministry of Finance and Public Credit, the National Institute of Statistics and Geography, the Tax Administration Service, and the Mexican Central Bank, referring to digital taxation.

**Keywords:** Digital taxation · Income tax · Tax collection

## 1 Introduction

In Mexico, the Tax System is compounded of three elements: income, expenses and debt. All expenses must necessarily have a counterpart in income, present or future, in such a way that the Income minus the Expenses is equal to the Debt or the Surplus. The fiscal incomes in Mexico are established in a Federal Income Law (LIF) and are obtained from the collection of taxes, that is, those that do not present consideration and are established with characteristics of coercion and imposition [1], see Table 1 and Table 2.

The important increases in tax collection after the Tax Reform of 2014, where the changes in bases, rates and objects, and can be seen in the federal government's income through the taxes that are in force. Figure 1 shows the main taxes collected in Mexico, in which you can see three of the main taxes for which tax revenues are obtained; such as: ISR, VAT and IEPS, the latter, due to the tax reform of 2014, went from having


M. d. P. Rodríguez García et al. (Eds.): XX SIGEF 2021, LNNS 384, pp. 82–93, 2022.
https://doi.org/10.1007/978-3-030-94485-8_7

**Table 1.** Percentage of income from taxes in Mexico.

| Income source | | Percentage |
|---|---|---|
| 1 | Taxes | 57% |
| 2 | Social security dues | 6% |
| 3 | Improvement contributions | 0% |
| 4 | Rights | 1% |
| 5 | Products | 0% |
| 6 | Other contributions | 1% |
| 7 | Income from sales of goods, provision of services and other income | 17% |
| 8 | Participations, contributions, agreements, incentives derived from tax collaboration and different contribution funds | 0% |
| 9 | Transfers, subsidies, gants, pensions and retirements | 9% |
| 10 | Income derived from financing | 9% |

Source: Own elaboration with data from SAT [2].

**Table 2.** Tax revenues in Mexico by type of tax

| | Income 2020 | % | Estimated income 2021 | % |
|---|---|---|---|---|
| Total | 6,107,732.40 | 100% | 6,295,736.20 | 100% |
| Taxes | 3,505,822.40 | 57% | 3,533,031.10 | 56% |
| Taxes over incomes: | | | | |
| Income tax | 1,852,852.30 | 30% | 1,908,813.40 | 30% |
| Patrimony tax | | | | |
| Taxes on production, consumption and transactions | | | | |
| Value added tax | 1,007,546.00 | 16% | 978,946.50 | 16% |
| Special production & consumption tax | 515,733.50 | 8% | 510,702.70 | 8% |
| New vehicle tax | 10,776.30 | 0% | 7,521.80 | 0% |
| Foreign trade tax | 70,984.60 | 1% | 61,638.40 | 1% |
| Payroll and assimilable taxes | | | | |
| Ecological taxes | | | | |
| Tax accessories: | 41,210.20 | 1% | 58,962.00 | 1% |
| Other taxes | 6,850.30 | 0% | 6,900.20 | 0% |
| Taxes not included in the LIF in force, caused in previous years pending settlement or payment | −130.80 | 0% | −453.9 | 0% |

Source: Own elaboration with data from SAT [2].

negative figures to positive figures, and according to the Economic Commission for Latin America and the Caribbean [2], Mexico managed to include tax sugary drinks, as well as non-essential foods with high caloric content, among others.

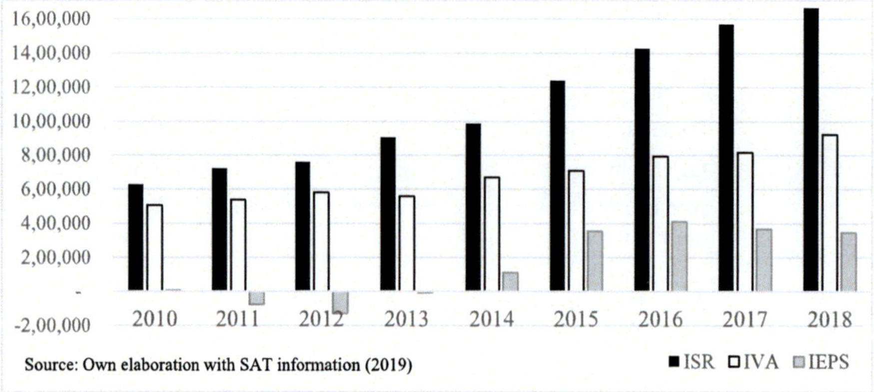


**Fig. 1.** Federal taxes in Mexico 2010–2018 (millions of pesos, MXN).

As can be seen in the previous graph, the main tax in Mexico is Income Tax (ISR), which at the end of 2018 represented 54% of the total tax collection of the federal government, followed by VAT, which represented 30%, and IEPS 11%, and 4% was collected from import tax, ISAN, IAEEH and other income.

The most important elements of taxes in Mexico, according to Osorio [4] are subject, object, base, rate or rate to determine a payment, either actively or passively, therefore, if any of these elements will reach Modified with the reforms, the determination of the tax would have a direct impact, as was the case of the IEPS mentioned above, in which its base was broadened to include sugary beverages, among other new objects.

Basically, there are classifications of taxes, where the first classifies them as direct, such as Income Tax and indirect, such as Value Added Tax and specials. At the same time, implement the digital tools and procedures used by the current tax administration, which promote digital taxation [5], which mainly consist of: increasing public collection, having greater control and administration of resources and guiding the taxpayer to make it is easier to meet your tax obligations.

## 2 Theorical Background

### 2.1 Incursion of Technologies in Mexico

Although it is true, the mechanisms in the fiscal issues in Mexico had a certain degree of obsolescence, however, as a result of the intensification of technologies since 2000, when technology was considered for the first time in the tax collection activity, it begins several changes and evolution of the way of paying taxes, according to Ratia, A., cited by Rodríguez, R. [6]. Among the changes in the tax mechanisms with the greatest impact in

Mexico, which intensified with the Tax Reform in 2014, are the adoption of the Electronic Bill (CFDI), the Electronic Signature, the Tax Mailbox, Referenced Tax Payments and Electronic Accounting [5].

The absence of technological tools to monitor effective compliance with tax obligations can generate a significant impact on the financial situation of organizations, which is why, in recent years, the SHCP and the SAT have evolved in terms of technology, capacity processing and databases. They are worried. These technological advances have made the transactions of the Authority and the taxpayers increasingly transparent. These changes have been called The New Tax Administration in Mexico [5].

Specifically, the 2014 Tax Reform intensified the use of information technologies in tax matters, however, it can be considered that said implementation began in 2000, due to the above based on the Development Plan of former President Vicente Fox.

Likewise, for Mexican legislation, the concept of Digital Taxation does not exist as such, however, according to the Costa Rican Ministry of Finance on its internet portal, it defines digital taxation as the following:

"Digital taxation is contextualized as a comprehensive tax management model based on electronic government strategies that will allow taxpayers to carry out their procedures, file returns, pay their taxes, resolve queries and comply with other tax duties through the internet" [7].

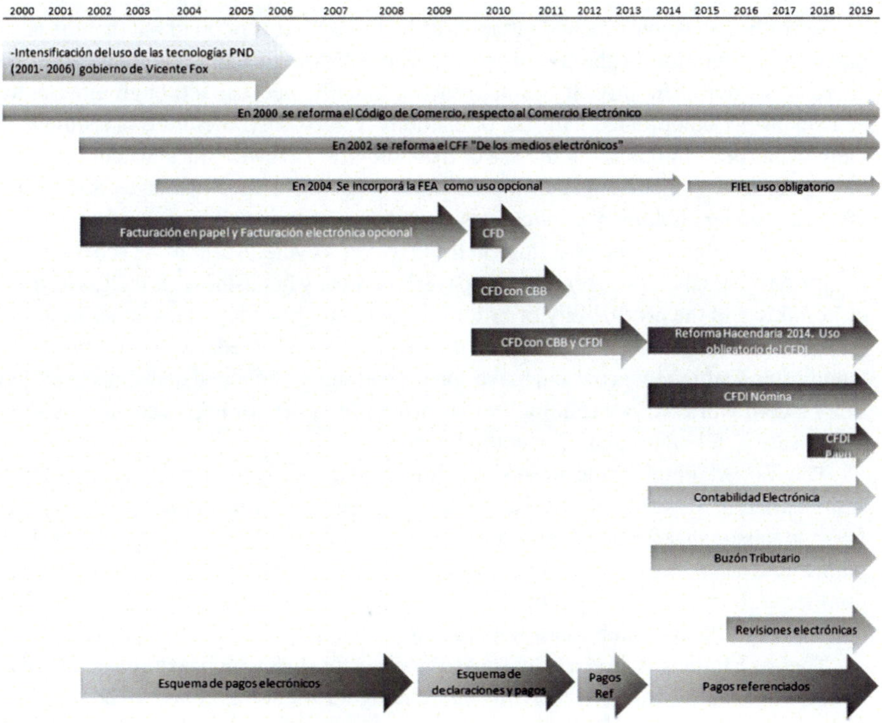


Source: Own elaboration with information from SAT [2].

**Fig. 2.** Timeline with the tools of SAT.

The previously defined concept is adapted to this research as part of the new tools developed by the Mexican Tax System, to be able to carry out procedures before the SAT (see Fig. 2).

One of the most remarkable advances for digital taxation in Mexico was the implementation of electronic invoicing, which began gradually from 2004 on an optional and mandatory basis in 2014, thus providing support and legality to operations [8].

Another example as part of the modernization, we can observe in the sending and filling of Declarations by Internet that the Service of the Mexican Tax Administration has used, specifically in the tool called Declarasat Online, which contains preloaded information of the taxpayer who enters to file his declaration, taking information from tax withholding agents, or Financial Institutions in the case of taxpayers who obtain interest income. The above allows the taxpayer to see preloaded information; Likewise, it allows you to review it and, if necessary, make the modifications that you consider pertinent, and later, using your electronic signature or your security keys, send your account statement at the same time.

Given the great modernization of economies and mechanisms, the SAT had the need to streamline the procedures that were carried out in person, through the Internet with the use of the Electronic Signature, which replaced the autograph signature. Today, the Electronic Signature allows more than 130 procedures to be carried out through the Internet, saving resources and travel time for taxpayers to the decentralized administrations of the SAT.

According to studies carried out in 2015 by the SAT, it is pointed out that the acceptance and integration of the use of information technologies and communication with citizens has not been easy; It has happened gradually, because although a tax culture of responsible compliance with tax obligations is necessary, a culture is required that demystifies perceptions about the use of technologies; Likewise, its constant use should be encouraged, which will help the taxpayer to confirm its advantages and become familiar with technological applications.

Keep accounting is the next stage in the taxpayer's cycle. Accounting provides information on the financial situation of a taxpayer's company or business, as well as its degree of liquidity and the profitability or profit it is generating, among many other aspects.

Electronic accounting is born from the taxpayer's own needs, as a solution to make their business efficient, grow and make correct and timely financial decisions [5]. Living in a modern world with technologies that make life easier for each person, this should be the case in all other areas, including financial.

The Tax Mailbox is the means that allows the taxpayer to interact with the tax administration. As of 2014, the Tax Mailbox is established in the tax provisions, as a free and personalized electronic communication system, located on the SAT electronic portal. It is a service that began its validity gradually, so that the taxpayer becomes familiar with its use. For legal persons, the obligation of use began on June 30, 2014, and for natural persons, on January 1$^{st}$, 2015.

The Tax Mailbox is an electronic post office box through which taxpayers can interact with the tax administration, receiving and sending information related to the fulfillment of their tax obligations. In this sense, communication is proposed in two ways: the issuing-receiving authority of the taxpayer and the issuing-receiving authority of the taxpayer, with reciprocity, in a continuous communication process [5].

The tax mailbox constitutes a new step in the technological transformation of the SAT whose purpose is to facilitate the spontaneous and timely fulfillment of tax obligations, which will reduce the response times of the authority regarding the matters handled by the taxpayer. Thus, the explanatory memorandum for the 2014 Tax Reform highlights the importance of migrating to this new interactive communication channel and exposes the effect that its use will have both for the tax administration, IMSS, INFONAVIT, and other agencies, as well as for the taxpayer.

For these purposes, the SAT mentions below that the personal notification process has a cost of 259 pesos per diligence, with an efficiency of 81%, and requires 5 to 30 days from the generation of the document to conclude the notification process. Therefore, using electronic and digital means for procedures will allow reducing, in addition to time and costs, the vices that currently exist, since they would be carried out immediately [5].

The Tax Mailbox is considered the only communication channel in which it is possible to interact and exchange, in real time, information, notifications, data and all kinds of documents with the SAT, and consequently the increase in the efficiency of the procedures, as well as saving time and money.

Electronic review is a procedure by which the authority identifies the compliance profile of taxpayers, as well as irregularities in terms of tax obligations.

According to the document Tax Administration in OECD Countries and Selected Non-OECD Countries: Comparative Information Series published by the OECD [9], in Mexico an average of 14 verification acts are carried out (audits and other types of reviews) per auditor in the year and only 0.6% of the universe of taxpayers is audited, unlike, for example, Chile, whose figures reach 251 and 4.6%, respectively.

One of the advantages of electronic reviews is that as a result of the permanent information analysis process carried out by the tax administration, the alleged irregularities that are identified can be corrected or clarified with greater opportunity, without necessarily resorting to audits of years or full periods. In this way, there will be an additional verification procedure that can avoid higher costs for the taxpayer and the authority.

The audit process not only aims to force compliance, but also to identify the causes that generate it, such as regulatory barriers and the lack of a service provided by the SAT that enables compliance, in this sense, the results of the audits They make it possible to specify the standard with the issuance of criteria, the development of services and, where appropriate, technological solutions to facilitate the correct fulfillment of tax obligations.

Precisely, the way to carry out the reviews through electronic means is part of a modernity objective anchored in the implementation of an electronic service provided by the authority, which allows the sending of digitized or electronic files on which the verification will be carried out. compliance.

## 2.2 Tax Collection in Mexico (Cobb-Douglas Model)

One of the ways in which the economic growth of a country can be projected is through the Cobb-Douglas production function. To represent the relationships between the product obtained, it uses the changes in the inputs capital (K) and labor (L), to which technology was later added, also called total factor productivity (TFP). It is a production function often used in economics. The origin of the Cobb Douglas function is found in the empirical observation of the distribution of total national income in the United States between capital and labor.

The Cobb-Douglas production function satisfies the statistical principles of standardized linear models where "a model must contain the least number of variables necessary to fit the data" [10]. In the present investigation, the Cobb-Douglas logarithmic production function is used as follows:

$$Y_t = AK^{\alpha}L^{\beta} \tag{1}$$

where: Y = Production, A = Productivity, and K y L = have values between 0 and 1 with constant returns.

According to Anderson et al. [11] and Sancho [12] testing the constant returns to scale hypothesis requires that the null hypothesis $H_0$: $\alpha + \beta = 1$ versus the alternative hypothesis $H_1$: $\alpha + \beta \neq 1$. Como $\alpha + \beta = 1$ implies that a = 1 − b for this, the restricted model is estimated [12] and the F test is used [11]. Now, this same function is used to represent tax collection, which was used by Carmona, et al. [13] to determine the main variables estimated by tax collection.

### 2.3 Theoretical Approach to the Research Problem

Due to the fact that in Mexico since 2000 innovations have been implemented and developed in the tax route, known in this research as Digital Taxation and that additionally the level of federal tax collection with respect to GDP only increased positively from 2014 to 2015, Mainly affected by the Tax Reform, which had important changes such as increases in rates and fees, increases in the tax bases due to reduction of deductions and exemptions, increases in the object of taxes for new taxable acts, which, is of sum It is important to analyze the true impact of digital taxation on federal tax collection in Mexico.

On the other hand, even with the modernization in the fiscal area, within the General Criteria of Economic Policy of 2019, the Federal Government does not contemplate a significant increase in the level of tax collection with respect to GDP higher than that of 2016.

Additionally, Mexico lacks scientific research that demonstrates that the implementation of digital taxation increases the proportion of federal taxes from 2014, however, for this study the year 2010 to 2020 is analyzed. For the above, we formulate the following hypothesis:

Hypothesis: *Digital taxation increases income tax collection in Mexico.*

## 3 Methodology

### 3.1 Data

The data referring to the Collection of Income Tax in Mexico on a monthly basis and during the years 2010 to 2020 were 132 observations obtained from the SHCP (https://

www.gob.mx) and the SAT (http://omawww.sat.gob.mx/cifras_sat). Figure 1 commented previously shows the Digital Taxation data for the measurement of the independent variables such as Electronic Bill (CFDI), Electronic Signature (FIEL), Referenced tax payments and number of active taxpayers. The SAT data is considered, in addition to the collection of macroeconomic data, such as the legal minimum wage in Mexico and the variation of the Global Economic Activity Indicator (IGAE) according to figures from INEGI's free database (https://www.inegi.org.mx), see Fig. 3.

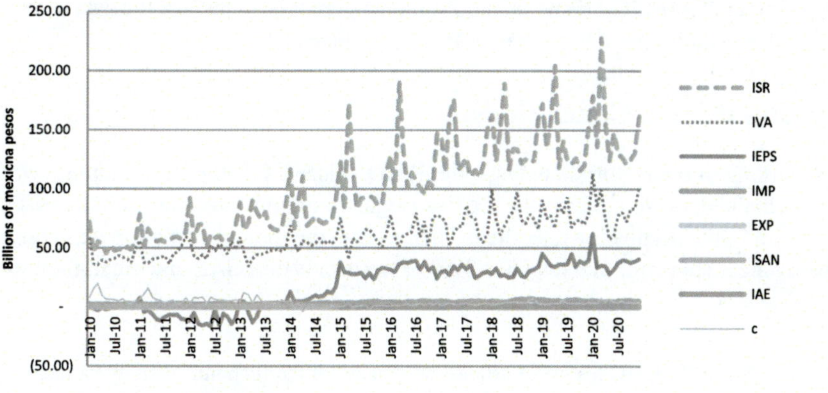


Source: Own elaboration with SHCP data.

**Fig. 3.** Collection of federal taxes in Mexico (2010 to 2020).

### 3.2 Econometric Model and Methods

Due to the nature of the data, it is carried out in the first instance in factorial analysis of macroeconomic data and according to [14], variables such as the variation of the IGAE and the Mexican legal minimum wage, consider the part of the economy as a control variable in the research. Once the economic part that affects income in Mexico has been established, a multiple linear regression is performed to determine the correlation of the Digital Taxation variables and their impact on Income Tax Collection.

Next, the model used to determine through multiple linear regression the impact of income tax collection derived from the implementation of digital taxation is shown.

$$Rec_t = \beta_0 + \beta_1 CFDI_t + \beta_2 TRT_t + \beta_3 EsIndT_t + \beta_4 EsLegT_t + \beta_5 ElectAcc_t + \beta_6 T_t + \beta_7 Eco_t + \varepsilon_t \quad (2)$$

where:
$Rec_t$ = Total federal tax collection in Mexico.
$CFDI_t$ = Natural logarithm of electronic invoice issued.
$TRT_t$ = Natural logarithm of total referenced tax payments.
$EsIndT_t$ = Natural logarithm of the electronic signatures for individual taxpayers.

$EsLegT_t$ = Natural logarithm of the electronic signatures for legal taxpayers.
$ElectAcc_t$ = Electronic Accounting.
$T_t$ = Taxpayers in force.
$Eco_t$ = Economic activity.
$e_t$ = Regression residuals.

An economic control variable was determined that includes the Minimum Wage of Mexico and the Global Indicator of Economic Activity in the SPSS statistical system in order to explain that the model is not only affected by digital taxation variables, but that there are also other macroeconomic variables or indicators. This combination, which includes the IGAE and the minimum wage, explains 89% of the data or explained variance, which were determined with a factor analysis.

## 4 Results and Conclusions

According to the factorial analysis, the KMO value of 0.500 of the economic variables was obtained, in this analysis the Mexican legal minimum wage, the monthly variation of the IGAE, the occupancy rate, the exchange rate, the monthly inflation was considered, being the main factor that explains 89% of the data is the IGAE and the minimum wage, see Table 3.

**Table 3.** KMO test and Bartlett's sphericity for the economic activity variable.

| Test | Statistical | Probability |
|---|---|---|
| KMO* | 0.500 | NA |
| Esfericidad de Bartlett | 120.371 | 0.0000 |

*Kaiser-Meyer-Olkin.

This economic activity was considered within the Multiple Linear Regression analysis with the independent variables of Digital Taxation and Income Tax Collection in Mexico, where it is obtained that the adoption of Digital Taxation directly benefits income tax collection, except for the issuance of Corporation Electronic signature (*EsLegT*), which reduces the collection of the main tax in Mexico. Additionally, the *ElectAcc* variable (fictitious/dichotomous) captures the mandatory nature of the Tax Mailbox and Electronic Accounting, which is significant. Table 4 shows the results.

The number of taxpayers is not significant, that is, increasing it does not generate greater collection. The model shows an adjusted R2 of 79.45%, which according to social science studies is acceptable according to Belady and Lehman [15]. The results of the previous table show that the implementation of digital taxation directly benefits income tax collection, except for the issuance of electronic signature certificates of legal persons, which seems to reduce the collection of the main tax in Mexico.

The model shows an adjusted $R^2$ of 79.45%, which according to studies in the social sciences is acceptable according to Belady and Lehman [15], this means that the best explanatory values of the variables of the income tax collection are correct and considerable.

**Table 4.** Results of the income tax collection model.

| Variable | Coefficient | Standard error | t statistic | Probability |
|---|---|---|---|---|
| Constant | 4.745373 | 1.156006 | 4.104973 | 0.0001 |
| *Eco* | 0.069348 | 0.033104 | 2.094844 | 0.0384 |
| *CFDI* | 0.152433 | 0.075560 | 2.017378 | 0.0460 |
| *TRT* | 0.285827 | 0.125706 | 2.273781 | 0.0249 |
| *EsIndT* | 0.140587 | 0.069516 | 2.022379 | 0.0455 |
| *EsLegT* | −0.375414 | 0.104726 | −3.584713 | 0.0005 |
| *ElectAcc* | 0.230515 | 0.072117 | 3.196413 | 0.0018 |
| T | 0.543225 | 0.355999 | 1.525919 | 0.1298 |

$R^2$ adjusted = 0.794597.
F statistic = 66.76426, probability (F statistic) = 0.0000.

Derived from technological advances in federal taxation and its implementation by the Tax Administration Service for taxpayers in Mexico in recent years, companies find themselves in the need to evolve towards the new digital era [5] named in this digital taxation investigation, which aims to have a positive impact on the collection of federal taxes.

Tax revenue is one of the most important sources for the State to provide public goods and services through taxes and other contributions. A recent study from 1990 to 2016 by the Tax Statistics in Latin America and the Caribbean [9], considers Mexico as one of the countries with the lowest tax collection in Latin America in proportion to average GDP of the OECD. With figures from the OECD, specifically the tax collection of the Federal Government with respect to GDP, Mexico presented a constant trend from 2000 to 2014, the year in which there was an important tax reform in tax matters, which promoted the use of technologies and implied important changes in its legislation, which increased to 12.8% of GDP and in 2016 tax collection reached its maximum point with 13.6%, a situation that did not continue in trend for the years 2017 and 2018, based on Mexican figures, it decreased from new.

It is true that Mexico has been one of the main countries in Latin America to implement digital taxation; however, no significant change has been observed in relation to GDP. The main tax income in Mexico is a direct tax called Income Tax with a proportion greater than 50% with respect to the total tax income obtained by the federal government, whose base is the generation of income, which in turn can be divide into gross income, benefit obtained and increase in equity.

So, to obtain these revenues, the Tax Administration in Mexico had been using traditional inspection tools such as information and payment declarations, the exercise of verification powers through traditional channels; However, given their obsolescence, they have begun to modernize, since the Tax Administration, day by day, acquires better technological tools to obtain the necessary information at the time of auditing, since technological progress has an exponential growth character and is highly exploitable in the tax area.

These new digital tools are the Tax Mailbox, Electronic Signature, Electronic Accounting, Referenced Tax Payments and mainly through Electronic Billing or CFDI [5], which in turn integrate the concept of Digital Taxation, which can be defined as a comprehensive tax management model based on electronic government strategies that will allow taxpayers to carry out their procedures, file returns, pay their taxes, resolve queries and comply with other tax duties through the internet.

Within the study carried out in this research, through secondary data published by the Ministry of Finance and Public Credit and the Tax Administration Service on tax revenues, as well as their possible causes or sources of origin such as the number of receipts issued, the number of certificates issued for the electronic signature of both natural and legal persons, mandatory tax mailbox, mandatory electronic accounting for taxpayers, referenced payments, 132 observations were made between the years 2010 to 2020, which in turn were subjected to normality tests, homoscedasticity, linearity and independence to have the correct data to perform the multiple linear regression.

Determining that the implementation of digital taxation directly increases income tax collection, the model shows an adjusted $R^2$ of 79.45%, which is acceptable according to social science studies such as Belady and Lehman [15]. Thus, the general objective of the research is fulfilled, which is to analyze the impact on income tax collection, derived from the implementation of the main Digital Taxation tools in Mexico.

## References


1. Sarur, M.S.: Ingresos no tributarios y el federalismo: asignación para el fondo de aportaciones múltiples. Ciencia Administrativa **1**, 86–99 (2014)
2. SAT: datos abiertos del Sistema de Administración Tributaria del Gobierno de México, http://omawww.sat.gob.mx/cifras_sat. Last accessed 11 May 2021
3. CEPAL: Panorama fiscal de América Latina y el Caribe. Naciones Unidas, Santiago de Chile (2017)
4. Osorio, J.M.: Análisis de la base gravable de los pagos provisionales del Impuesto Sobre la Renta de las personas Morales en México. Doctoral Thesis, Instituto Especializado para Ejecutivos, Guadalajara (2013)
5. SAT: La nueva administración tributaria en México: El ADN digital: eje de transformación de los servicios tributarios. Servicio de Administración Tributaria, Ciudad de México (2015)
6. Rodríguez, J.C.: La nueva hacienda pública distributiva en México, Año 2001, rechazada por el Congreso de la Unión. Master Thesis, Universidad de Occidente (2009)
7. MHCR Homepage: https://tributaciondigital.hacienda.go.cr/irj/portal/anonymous. Last accessed 6 Nov. 2018
8. De León, V., Cerón, M.T., León, F.J., Rodríguez, S.: Impacto de la implementación de la factura electrónica en las MiPyMes del sector comercio y servicios. Rev. Global Negocios **4**(7), 85–94 (2016)
9. OECD: Revenue Statistics in Latin America and the Caribbean 1990–2016. OECD Publishing, Paris (2018)
10. Navidi, W.: Estadística para ingenieros. 1st edn. (Spanish version). McGraw Hill, Ciudad de México (2006)
11. Anderson, D., Sweeney, D., Williams, T.: Estadística para administración y economía, 10th edn. Cengage Learning Editores, Ciudad de México (2008)
12. Sancho, A.: Econometría de económicas, función de producción Cobb-Douglas. https://www.uv.es/sancho/funcion%20cobb%20douglas.pdf. Last accessed 11 May 2021

13. Carmona, A., Molina, A., Ruíz, A.: Determinantes del ingreso tributario en México. Anál. Econ. **34**(87), 177–197 (2019)
14. CEPAL: Panorama fiscal de América Latina y el Caribe. Naciones Unidas, Santiago de Chile (2019)
15. Belady, L.A., Lehman, M.M.: The characteristics of large systems. In: Wegner, P. (ed.) Research Directions in Software Technology. The MIT Press, Cambridge, MA (1979)

# Methodological Issues

# Analysis of Poverty Through Educational Lag Using the Maximum Clique into the Complex

Israel Santiago-Rubio[1], Román Mora-Gutiérrez[2], Edwin Montes Orozco[2](✉), Eric Alfredo Rincón García[3], Sergio Gerardo de los Cobos Silva[4], Pedro Lara Velazquez[4], and Miguel Ángel Gutiérrez Andrade[4]

[1] Maestría en Ciencias de la Computación, Universidad Autónoma Metropolitana, Azcapotzalco, Av. San Pablo Xalpa 180, Col. Reynosa Tamaulipas, Azcapotzalco, CP 02200, México

[2] Departamento de Sistemas, Universidad Autónoma Metropolitana, Azcapotzalco. Av. San Pablo Xalpa 180, Col. Reynosa Tamaulipas, Azcapotzalco, CP 02200, México
emontes@azc.uam.mx

[3] Iztapalapa San Rafael Atlixco No. 186, Col. Vicentina, Iztapalapa, CP 09340, México

[4] Departamento de Ingeniería eléctrica, Universidad Autónoma Metropolitana, Iztapalapa San Rafael Atlixco No. 186, Col. Vicentina, Iztapalapa, CP 09340, México

**Abstract.** In this work, it is proposed to detect communities that share attributes in common, based on the attributes that quantify the indicator of deficiency due to educational lag, which is proposed by the National Council for the Evaluation of Social Development Policy (CONEVAL) in its multidimensional poverty analysis methodology in Mexico. The data obtained to carry out this work were recovered from the databases generated by the National Institute of Statistics and Geography (INEGI) from the National Survey of Household Income and Expenditure in 2018 (ENIGH-2018). The proposed methodology consists of 1) recovering the data of the indicator of lack due to educational lag, 2) a characterization was carried out on each of the variables by states of the Mexican Republic, 3) a complex network model was generated for the indicator of deficiency analysed, 4) the complex network was analysed using a genetic algorithm to detect the maximum click in the network, with this it was possible to detect communities made up of federative entities with similar attributes 5) the properties of the subgraph of statistical form. The objective of the proposed methodology aims to address the analysis of multidimensional poverty in Mexico from a complex network and optimization approach.

**Keywords:** Social network · Community analysis · Data analysis

## 1 Introduction

In the present work, the poverty problem is analyzed from a complex network perspective and optimization. For this, complex networks were built based on well-being, social


M. d. P. Rodríguez García et al. (Eds.): XX SIGEF 2021, LNNS 384, pp. 97–108, 2022.
https://doi.org/10.1007/978-3-030-94485-8_8

rights indicators in Mexico. To detect communities with similar attributes into networks generates the problem of the maximum clique (PMC) was used.

The within the network is classified as an NP-hard problem. The approach proposed in this work consists of analyzing the network models using a genetic algorithm, which, due to its characteristics, provides reasonable solutions to the problem posed.

The present work is structured as follows: The first section presents basic concepts. Section 2 presents a brief review of the overview. Next, in Sect. 3, the methodology used is described. Then in Sect. 4, the results obtained are shown. Finally, in Sect. 5, the conclusions are presented.

## 2 Basic Concepts

### 2.1 Complex Networks

A concise definition of a complex system there is not in the literature. However, there are a set of characteristics common by complex systems, which are non-linear dynamic, feedback, self-organization, robustness, and emergence [1, 2], and [3]. Internet, propagation of COVID-19, the cell, the human brain, or society, among others, are some examples of a complex system.

A system complex can be represented and analyzed through the network. In this representation, elements of the system (or agents or comports) are represented by nodes. On the other hand, interactions or relationships among nodes are drawn by links [4]. These objects are studied and analyzed by network science, an interdisciplinary field of science [5], and [6] that involves both ideas and knowledge physic, mathematic, sociology, and economics. Several works of the network aim to understand the structural properties of real complex networks and introduce proper models to construct synthetic networks mimicking their properties.

Complex networks have been used in biology for modeling: disease transmission networks, ecological networks, metabolic networks, genetic networks, among others. Any other way, complex networks have been used for representing distribution networks, telecommunications, military operations, transport networks, internet networks, among others. Finally, they are used in social sciences by draw affiliation networks, broadcast networks, groups of persons, and communication networks.

Formally, a complex network can be defined as an undirected (directed) graph $G = (\mathcal{N}, \mathcal{L})$ consists of two sets $\mathcal{N}$ and $\mathcal{L}$, such that $\mathcal{N} \neq \varnothing$ and $\mathcal{L}$ is a set of unordered (ordered) pairs of elements of $\mathcal{N}$. The elements of $\mathcal{N} \equiv \{n_1, n_2, ..., n_N\}$ are the nodes (or vertices, or points) of the graph $G$, while the elements of $\mathcal{L} \equiv \{l_1, l_2, ..., l_K\}$ are its links (or edges, or lines). N and L elements are denoted by $N$ and $K$, respectively [7]. The set of metrics and characteristics more usually determined in complex networks are represented in Table 1.

Frequently, nodes of the complex network cluster together into distinct groups. Properties of these groups are independent of the properties of individual nodes. These groups are known as communities, and this kind of network structure is known as community structure [8].

**Table 1.** Metrics and characteristics of complex networks

| Characteristic | Symbol | Equations |
|---|---|---|
| Degree or connectivity | $k_i$ = is the number of edges incident in $i-$ th node | $k_i = \sum_{j \in \mathcal{N}} a_{i,j}$ where $a_{i,j}$ is the time in the $i - th$ row and $j - th$ column of the adjacency matrix |
| Average degree | $k$ = is the sum of all the inputs of the adjacency matrix divided by the number of nodes | $k = \frac{1}{N} \sum_{i=1}^{N} k_i$ |
| Grade distribution | $P(k)$ = is the probability that a particular node chosen at random has degree k | Poisson topology $P(k) = e^{\lambda} \frac{\lambda}{K!}$ Exponential topology $P(k) = Ce^{-\alpha k}$ Power Law Topology $P(k) = Ck^{-\gamma}$ |
| Assortative coefficient | $r_k$ = represents the tendency of nodes to establish links with others similar to them | $r_k = \frac{\frac{1}{M}\sum_{i,j} k_i k_j - \{\frac{1}{M}\sum_{i,j}\frac{1}{2}(k_i+k_j)\}^2}{\frac{1}{M}\sum_{i,j}\frac{1}{2}\left(k_i^2+k_j^2\right) - \left\{\frac{1}{M}\sum_{i,j}\frac{1}{2}(k_i+k_j)\right\}}$ |
| Transitivity | $T$ = measures the global grouping of the network | $T = \frac{numberoftrianglesinthenetwork}{numberofsetsofthreevertices}$ |
| Clustering coefficient | $C_i$ = is a measure of the degree to which nodes in a graph tend to cluster together | $C_i = \frac{2T(i)}{k_i(k_i-1)}$ |
| Average clustering coefficient | $C$ = is the average of the local clustering coefficients of all the vertices | $\overline{C}_i = \frac{1}{N} \sum_i C_i$ |

A clique is a subset of vertices of a graph such that every two distinct vertices in the clique are adjacent. In other words, a clique $G(C)$ is a sub graphic complete of a graphic $G$.

## 2.2 Poverty and Educational Lag

The CONEVAL (National Council for the Evaluation of Social Development Policy) mentions that a human is in a situation of multidimensional poverty when the exercise of at least one of his rights for social development is not guaranteed, and his income is insufficient to acquire the goods and services he requires to satisfy his needs [9]. The dimensions considered in the poverty analysis are; the economic well-being dimension, the social rights dimension, and the territorial scope. Each one provides a diagnosis of the limitations and restrictions that people face.

Although the presence of deficiencies associated with each of the spaces imposes a series of specific limitations that threaten the freedom and dignity of people, the simultaneous existence of deficiencies in the spaces considerably aggravates their living conditions, which gives rise to the following definition of multidimensional poverty.

## 3 Overview

### 3.1 Studies of Communities into a Complex Network

Developing a method that can evaluate the existing closeness between nodes results in a highly efficient tool to explore the community structure in real networks. The problem of finding close neighbors is analyzed in [10, 11], for which the graph is transformed into a metric space, and the metric tree (M-tree) and the locality-sensitive hash (LSH) are determined. In [12], Newman proposes to detect communities using matrices of eigenvectors considering community detection with modularity in a complex network. On the other hand, Pizzuti et al. the GA-NET, a genetic algorithm in social networks that uses a fitness function with the aim detected modularity of the network, is presented [13]. Gong et al. in [14] proposed a MEMETIC-NET, an algorithm for community detection optimizing based on the modularity density function. Ritter et al. [15] developed a cluster detection algorithm based on close neighbors identifying communities with different densities. Ferreira and Zhao [16] presented a complex network-based time-series cluster technique to detect communities, effectively solving the general task. Mahmood and Small [16] time series analysis. In [17], an improved method of detecting communities on the basis that community detection by maximizing modularity is presented. Xin et al. mentioned a dynamic community detection method based on a random sample walk, which considers people's tendency to connect with friends [18]. In [19], an algorithm is proposed to carry out community detection based on local similarity and grade grouping information.

Cheng et al. proposed the LMOEA algorithm, based on the research of the multiobjective evolutionary algorithm, which is used for community detection using information from the local community [20]. Mohammed et al. [21] developed a variance-based differential evolution algorithm with an optional crossover for data clustering to improve the quality of clustering solutions and convergence speed. Alireza Moayedikia [22] proposes a novel algorithm multiobjective to detect attributed community based on node importance analysis; it solves community detection and does not consider interactions between nodes. In [23], a measure to numerically evaluate the community robustness of interdependent networks is designed; and proposes a memetic optimization algorithm. Feng et al. to the [24] a novel multiobjective discrete backtracking search optimization algorithm with decomposition to detect communities in complex networks is presented.

### 3.2 Educational Lag in Mexico (2016–2021)

In [25], a review of historical interrelations and educational policy, which make up the identity reference of the curriculum of Mexican primary education, is presented. In [26] reported an analysis about what theories, policies, and environment educational related-contents were included in the documents that integrated this curriculum: general study plan, study programs of Grades and official students' textbooks.

In [27], the augmented reality technology has a positive impact on learning-related outcomes of middle-school Mexican students is shown. In [28], the relationship between PISA 2012 maths test scores and relative poverty is studied.

### 3.3 Problem Statement

Currently, organizations such as the OUN, UNESCO, OECD, FAO have focused their interest on economic inequality and poverty. New poverty indicators and analysis techniques such as the Human Development Index (HDI) have shown their limitations as they are based on linear models.

In Mexico, the concept of educational lag is used to refer to people over 15 years of age who have not completed secondary education. The INEA (National Institute for Adult Education) mentions that in 2019 there were a total of 28.6 million people with educational backwardness, that is, 30.6% of the total of people over 15 years of age [27]. At present, there is the challenge of designing modern indices that consider the non-linear relationships between the various variables and working with large volumes of data. In this work, a strategy for analyzing the educational poverty is presented. It is based on science network and optimization.

## 4 Methodology

This section describes the steps employed to analyze poverty education in Mexico. The process used is shown in Fig. 1.

First, we obtained the data collection for the generation of poverty indicators, we considering the databases from the 2018 Statistical Model were used for the continuity of the MCS-ENIGH of INEGI. The second step is the treatment and structuring of information, for that, the data set was cleaned, homogenized and structured in tables for handling. In this phase the R program was used. later, the information generated by state was concentrated and the basic descriptive statistics were obtained. The third is generate network model, we are constructing complex network models, the Mahalanobis distance measure was applied to the M1 matrix, which allows determining the similarity between two multidimensional random variables. The fourth step is to detect maximum clique, to detect groups of states that share similar characteristics, a genetic algorithm was configured in the MatLab environment [29], with which the network models obtained were analyzed. The input parameters that the Genetic algorithm requires are an initial population of 10 individuals, iterate 500 times, a survival rate of 0.30%, and the M3 matrix. With these configurations, the algorithm returns a matrix of M5 solutions with 20 approximations per indicator. Final, we analyze the groups obtained, this process returns statistical summaries that show the groups of Entities that share common characteristics, such as the national average by indicator and the entities that do not share common attributes.

## 5 Experimental Results

The historical information since 2008 contained in the National Survey of Household Income and Expenditure New Series (ENIGH-NS) was obtained from INEGI, which underwent a data processing phase to characterize the educational backwardness indicator by state. The network model was generated based on the previous information. This model is shown in Fig. 1. On the other hand, the characteristic structural educational lag Network Model.

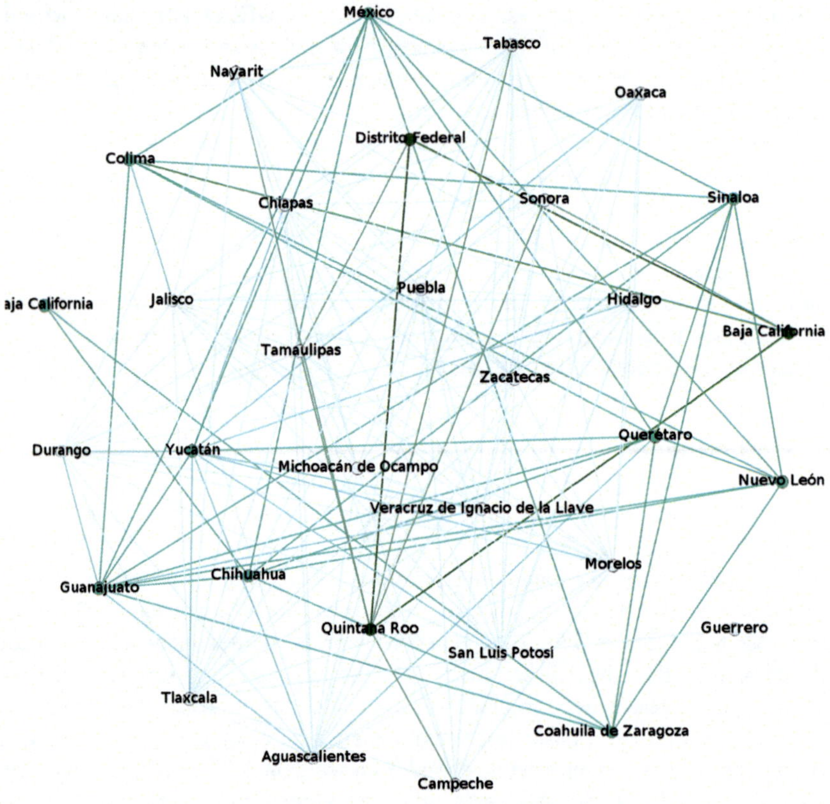


**Fig. 1.** Educational lag network model

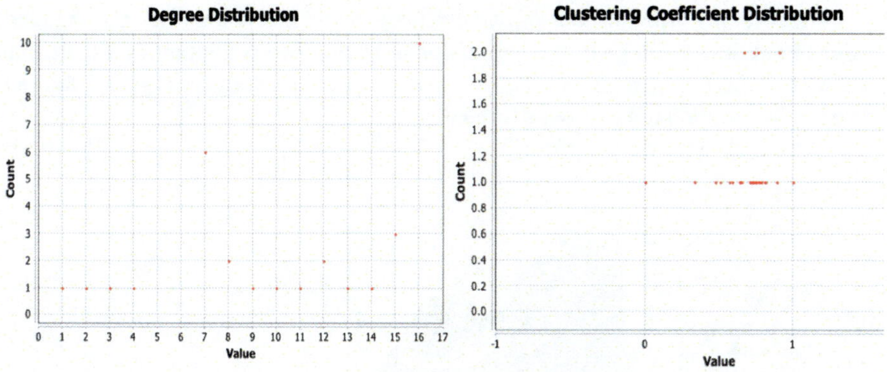


**Fig. 2.** Characteristic structural educational lag network model

Subsequently, the network is analyzed in order to find the maximum click. To solve the maximum clique problem through a genetic algorithm with 20 repetitions was applied (see Fig. 2).

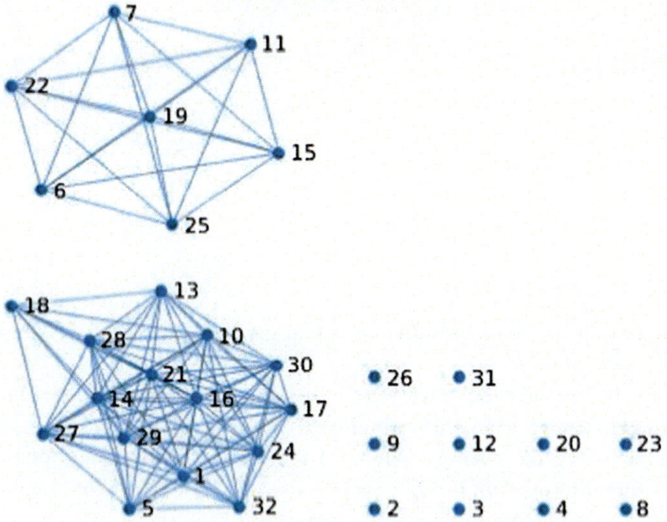


**Fig. 3.** Result per maximum click for Educational Lag

The resulting grouping was of two clusters, one with seven entities and the second with 15 entities, which gives the impression of having two communities. Figure 3 also shows that there are ten entities isolated to the two groups formed. Figure 4 shows the geographic location of the groups formed.

**Fig. 4.** Geographic location for Educational Lag

The box diagram shown in Fig. 5 corresponds to Educational Lag. It describes the entities that make up the maximum clique in a general way. In the case of this indicator, two groups were formed.

After analyzing the average values of the indicators of educational lag of group one, the information shown in Fig. 6 is obtained.

After analyzing the average values of the indicators of educational lag of group two, the information shown in Fig. 7 is obtained and the Fig. 8 for the similarity of the educational lag indicators for each federal entity.

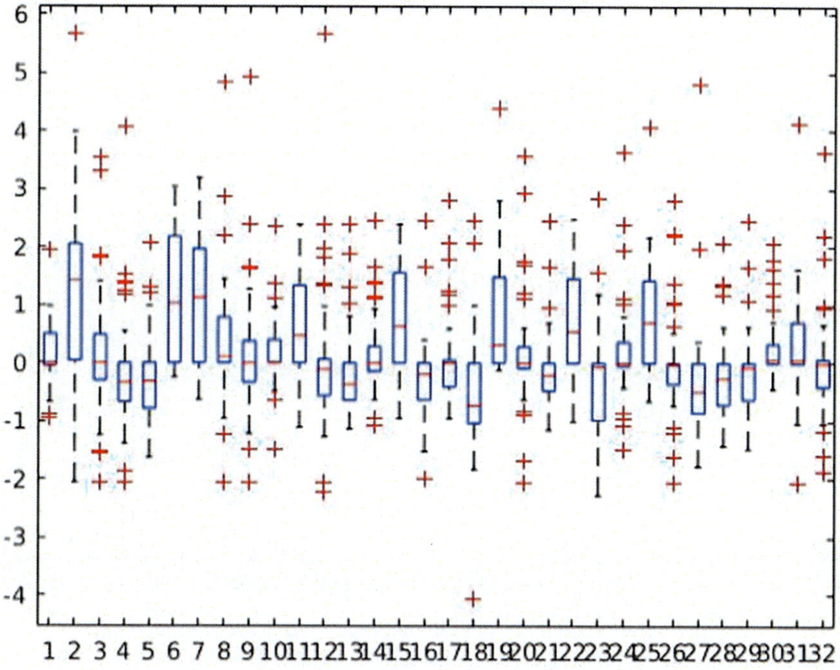


**Fig. 5.** Comparison by federal entities for Educational Lag

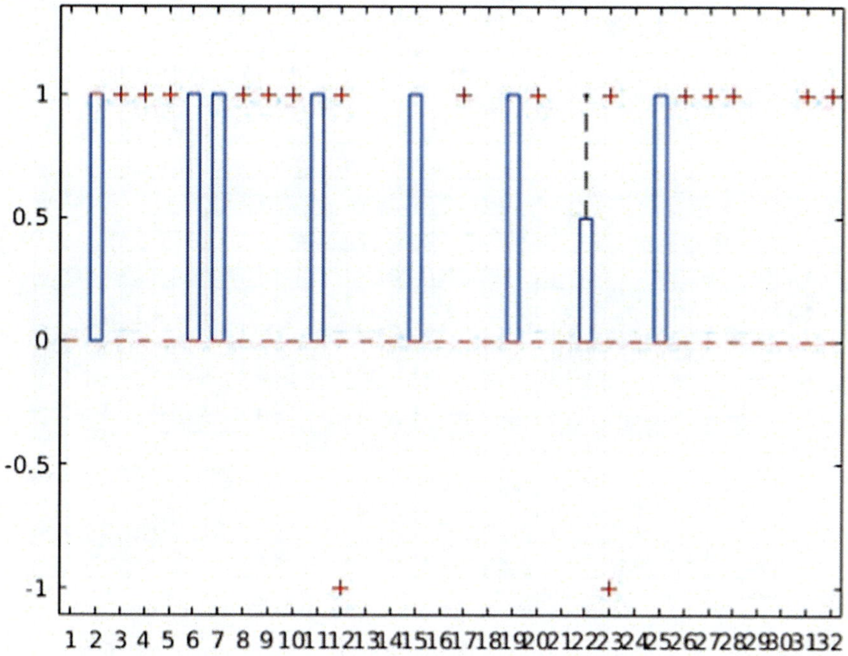


**Fig. 6.** Comparison of values of the first group for Educational Lag

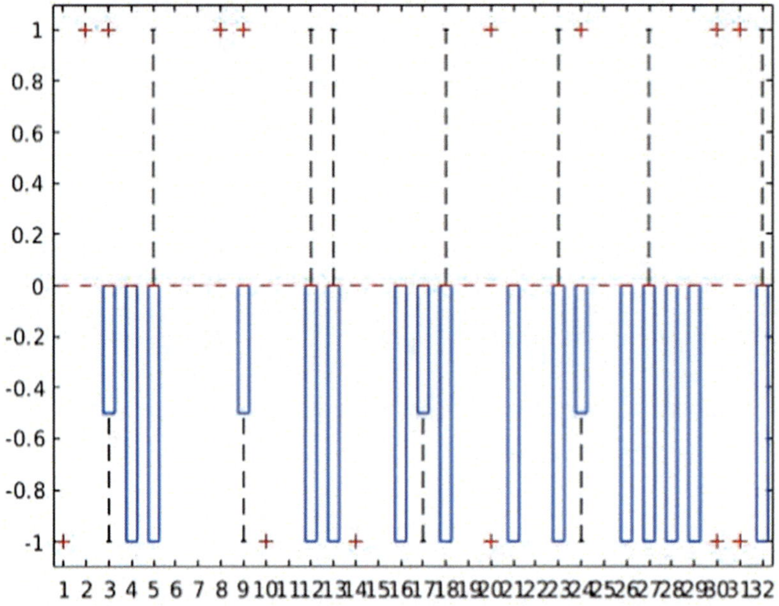

**Fig. 7.** Comparison of values of the second group for Educational Lag

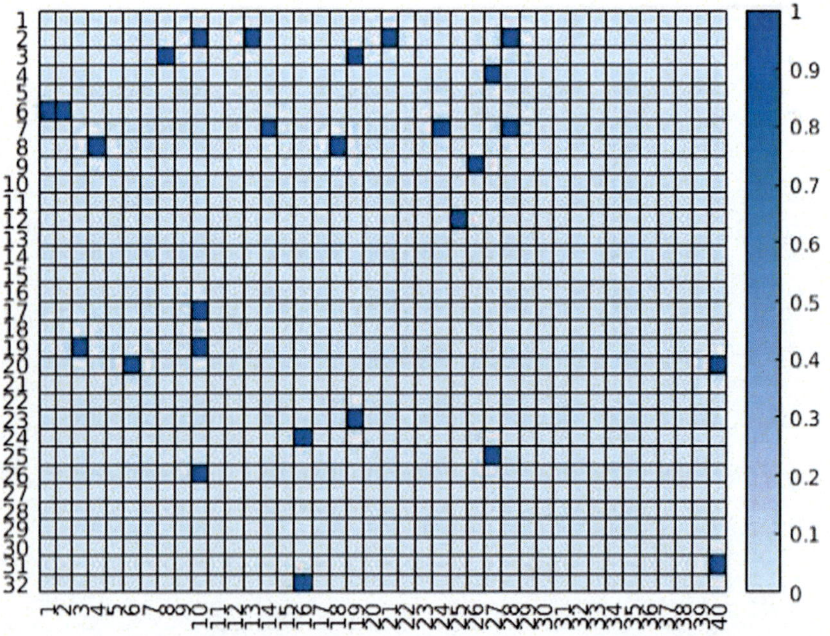

**Fig. 8.** Relationship of similarity of the educational lag indicators for each federal entity

## 6 Conclusions

The results obtained show the existence of two blocks in Mexico due to educational backlog. The first group formed by Sinaloa, Coahuila, Chihuahua, Nuevo Leon, Guanajuato, Queretaro and the State of Mexico have similar attributes. On the other hand, the second group formed by Durango, Zacatecas, San Luis Potosi, Nayarit, Jalisco, Michoacán, Hidalgo, Veracruz, Tlaxcala, Puebla, Chiapas and Tabasco have similar attributes.

Both groups show the typical characteristics of the educational backwardness in Mexico. The states not grouped in any of the groups indicates that it has little similarity with the members of the groups; for example, Oaxaca and Guerrero that have shallow values for all indicators of educational backwardness are outside of both groups. In contrast, the CDMX with high values in the indicators is also outside both groups.

## References


1. Paradisi, P., Kaniadakis, G., Scarfone, A.: The emergence of self-organization in complex systems-Preface (2015)
2. Jacobson, M., Levin, J., Kapur, M.: Education as a complex system: Conceptual and methodological implications. Educ. Res. E **48**(2), 112–119 (2019)
3. Haken, H., Portugali, J.: Information and self-organization. Entropy E **23**(6), 707 (2017)
4. Amaral, L., Ottino, J.: Complex networks: augmenting the framework for the study of complex systems. Eur. Phys. J. B E **38**(2), 147–162 (2004)
5. Börner, V., Sanyal, S., Vespignani, A.: Network science. Ann. Rev. Inf. Sci. Technol. E **41**(1), 537–607 (2007)
6. Perc, M., Jalili, P.: Information cascades in complex networks. J. Complex Netw. E **5**(5), 665–693 (2017)
7. Boccaletti, S., Latora, V., Chavez, M., Moreno, Y.: Complex networks: structure and dynamics. Phys. Rep. E **424**(4–5), 175–308 (2006)
8. Fang, K., Sivakumar, B., Woldemeskel, F.: Complex networks, community structure, and catchment classification in a large-scale river basil. J. Hydrol. E **545**, 478–493 (2017)
9. CONEVAL: Medición multidimensional de la pobreza en México. 3rd edn. Consejo Nacional de Evaluación de la Política de Desarrollo Social, Ciudad de México (2019)
10. Saha, S., Ghrera, S.P.: Nearest neighbor search in the metric space of a complex network for community detection. Information **7**(1), 17 (2016)
11. Newman, M.: Finding community structure in networks using the eigenvectors of matrices. Phys. Rev. E **74**(3), 036104 (2006)
12. Pizzuti, C.: Ga-net: a genetic algorithm for community detection in social networks. In: Rudolph, G., Jansen, T., Beume, N., Lucas, S., Poloni, C. (eds.) PPSN 2008. LNCS, vol. 5199, pp. 1081–1090. Springer, Heidelberg (2008). https://doi.org/10.1007/978-3-540-87700-4_107
13. Gong, M., Fu, B., Jiao, L., Du, H.: Memetic algorithm for community detection in networks. Phys. Rev. E **84**(5), 056101 (2011)
14. Ritter, G., Nieves, J., Urcid, G.: A simple statistics-based nearest neighbor cluster detection algorithm. Pattern Recogn. E **48**(3), 918–932 (2015)
15. Ferreira, L., Zhao, L.: Time-series clustering via community detection in networks. Inf. Sci. E **326**, 227–242 (2016)
16. Mahmood, A., Small, M.: Subspace based network community detection using sparse linear coding. IEEE Trans. Knowl. Data Eng. E **28**(3), 801–812 (2015)

17. Xin, Y., Xie, Z., Yang, J.: An adaptive random walk sampling method on dynamic community detection. Expert Syst. Appl. E **58**, 10–19 (2016)
18. Wang, T., Yin, L., Wang, X.: A community detection method based on local similarity and degree clustering information. Physica A Stat. Mech. Appl. E **490**, 1344–1354 (2018)
19. Cheng, F., Cui, T., Su, Y., Niu, Y., Zhang, X.: A local information based multi-objective evolutionary algorithm for community detection in complex networks. Appl. Soft Comput. E **69**, 357–367 (2018)
20. Alswaitti, M., Albughdadi, M., Isa, N.: Variance-based differential evolution algorithm with an optional crossover for data clustering. Appl. Soft Comput. E **80**, 1–17 (2019)
21. Moayedikia, A.: Multi-objective community detection algorithm with node importance analysis in attributed networks. Appl. Soft Comput. E **67**, 434–451 (2018)
22. Wang, S., Liu, J.: Community robustness and its enhancement in interdependent networks. Appl. Soft Comput. E **77**, 665–677 (2019)
23. Zou, F., Chen, D., Li, S., Lu, R., Lin, M.: Community detection in complex networks: Multi-objective discrete backtracking search optimization algorithm with decomposition. Appl. Soft Comput. E **53**, 285–295 (2017)
24. Gutiérrez, A.: The identity reference of the Mexican basic education curriculum: causes and consequences of curricular injustice. Educação Unisinos E **22**(1), 74 (2018)
25. Paredes, A., Viga, M.: Environmental education (EE) policy and content of the contemporary (2009–2017) Mexican national curriculum for primary schools. Environ. Educ. Res. E **24**(4), 564–580 (2018)
26. Ibáñez, M., Uriarte, A., Zatarain, R., Barrón, M.: Impact of augmented reality technology on academic achievement and motivation of students from public and private Mexican schools: a case study in a middle-school geometry course. Comput. Educ. E **145**, 103734 (2020)
27. Daniele, V.: Socioeconomic inequality and regional disparities in educational achievement: the role of relative poverty. Intelligence E **84**, 101515 (2021)
28. Mendoza, M.: Rezago social y letalidad en México en el contexto de la pandemia de enfermedad por coronavirus (COVID-19): una aproximación desde la perspectiva de la salud colectiva en los ámbitos nacional, estatal y municipal. Notas de Población (2021)
29. MATLAB: R2017a. The MathWorks Inc., Natick (2021)

# Spatial Effects on Economic Convergence Among Mexican Firms

Esteban Picazzo Palencia(✉), Jeyle Ortiz Rodríguez, and Elias Alvarado Lagunas

Universidad Autónoma de Nuevo León, San Nicolas de los Garza, Nuevo León, México
esteban.picazzopln@uanl.edu.mx

**Abstract.** After economic openness Mexico experienced a series of changes in its economic structure. These events led to changes in economic growth of cities, which affected firms' decision on location. Using the developed model to measure conditional convergence and including the effect of geographic context, this paper explores the existence or lack of a process of convergence among the economic production of Mexican firms in municipalities from 2013 and 2018. Results reveal that from 2008 to 2013 firms' production has experienced a process of divergence, and location determines the growth of firms' output. The hypothesis that geographical location is a determinant of growth is supported. This suggests that location in regions with high levels of firms' production positively affects the growth of production, and firms located in municipalities with higher levels of production reported greater growth than firms in municipalities with lower firms' production levels. Also, the results of this study show that location determines the successful of firms. This indicates that firms' production increases based on cluster.

**Keywords:** Convergence · Firms' production · Spatial analysis

## 1 Introduction

One of the main concerns of any country is economic growth. The rationale for many policies and legislation is economic growth that promises to guarantee citizens an adequate standard of living. Governments often implement new or radical policies in order to reach this goal. However, these policies do not always have the expected effect; sometimes these policies can have the opposite effect leading to severe crisis and public debt. Determining the effectiveness of these policies can be difficult when regions within countries grew differently. Although the results of some policies can be unfavorable for some people, their objective is to increase individuals' living standards and create equal opportunities for people from different socioeconomic status.

Mexico is notorious for the high levels of inequality among its different regions. For instance, in 2019 the Gross Domestic Product (GDP) of Distrito Federal and Nuevo Leon was 3,139,561 and 1,377,526 million pesos respectively, while the GDP of Tlaxcala and Colima was 104,653 and 109,496 million pesos [1]. Convergence among regions is difficult when there are differences in the economic growth rates between regions over


M. d. P. Rodríguez García et al. (Eds.): XX SIGEF 2021, LNNS 384, pp. 109–121, 2022.
https://doi.org/10.1007/978-3-030-94485-8_9

time –more specifically, when already wealthy regions have higher growth rates than economically depressed regions. Regions with larger economic growth will accumulate more capital and qualified personnel, which will increase their profitability relative to the more disadvantaged regions. Therefore, there will be fewer incentives to invest in regions with low economic development making economic integration of poorer regions less likely.

Convergence among regions is difficult when differences in economic growth between regions over time exist. Regions with larger economic growth will accumulate more capital and qualified personnel, which will increase their productivity compared to disadvantaged regions. Therefore, there will be fewer incentives to invest in regions with low rates of economic development causing economic integration of poorer regions less likely.

Some economists [2–4], argue that convergence between rich and poor regions of Mexico started before 1985. However, evidence of convergence disappeared after Mexico signed the General Agreement on Tariffs and Trade (GATT) in 1986. In addition, the reduction in convergence appears to have intensified after the North American Free Trade Agreement (NAFTA) was enacted in 1994. These agreements liberalized trade policies and opened Mexico's borders promising economic growth, but they only have a positive impact on rich regions reducing the convergence process and increasing economic inequality among regions. New economic theories include the importance of spatial effects, where clusters located in geographic areas take advantages of aspects related to the geographic context producing benefits for local firms [5]. There is a lack of studies that explore the effect of geographic aspects on economic integration in Mexico.

## 2 Objective

The objective of this study is to analyze the convergence process in the production level of Mexican firms from 2013 and 2018. Using the well-known model to measure conditional convergence we include the effect of geographic contexts.

## 3 Theoretical Framework

Regarding economic convergence, Chiquiar [4] finds out that before trade liberation, regions in Mexico experienced a positive convergence velocity, but after Mexico entered the GATT in 1986, economic divergence started to appear. Chiquiar [4] associates the divergence process of economic development to differences in the existing infrastructure before economic openness, which allowed some regions to develop advantages. The author identifies that after NAFTA, enterprises that initially were located in Mexico City moved to the northern border of Mexico in order to reduce transportation costs. This increased the education level in the workforce of that region.

Similarly, Arroyo [2] analyzed the convergence process of Mexican states and finds out that economic growth experienced convergence from 1980 to 1985 and divergence from 1985 to 1999. He highlights that the decreasing intervention of the State contributed to intensify the gap in the production among Mexican regions.

Although many authors argue the existence of a divergence process of economic activities in Mexico, a few explore the role of geographic context in the economic integration of firms from different municipalities. According to López [6], dominant economic activities create a multiplier effect on the nearest firms. Basically, growing industries with strong economic power may promote the integration of small businesses to economy. The integration of different firms is translated into clusters, which contributes to the development of a profitable economic region.

Porter [5] defines a cluster as a group of related firms (horizontal or vertical) located in a geographic area that benefit from a series of aspects, such as low transportation costs, availability of inputs, specialized workforce. Based on Porter's model, clusters promote firms' competitiveness through three channels: they i) increase firms' productivity; ii) improve their capacity to innovate; and iii) promote the creation of new firms that promote innovation.

Within a country, regions have different initial conditions in terms of infrastructure and human capital. Differences in existing conditions may produce dissimilar economic growth Barro and Sala-i-Martin [7]. Regions with more favorable conditions will attract higher levels of investment. Meanwhile, firms will have lower incentives to locate in regions with poor infrastructure and a low concentration of firms. This concept if called conditional convergence.

## 4 Methods

Barro and Sala-i-Martin (1992) developed the following model to measure absolute convergence:

$$\frac{1}{T}\log\left[\frac{y_i, t_{o+T}}{y_{i,t_0}}\right] = \alpha - \frac{1-e^{-\beta T}}{T}\log\left[y_{i,t_0}\right] + u_{i,t_0 \cdot t_0+T} \tag{1}$$

where:

$y\_(i, t\_0)$ is the initial level of per-capita production of firms located in municipality $i$.

$y\_(i, t\_0 + T)$ is the final level of per-capita production of firms located in municipality $i$ after T years.

$u\_(i, t\_0 \cdot t\_0 + T)$ is a distributed lag of the residuals.

$$\alpha = x + \left(1 - e^{(} - \beta T)\right)/T \quad log\left[y*\right] + x\_(t\_0)$$

This model requires all municipalities to have the same technological progress level and steady state per-capita level. This assumption is very strong given the heterogeneity of Mexican municipalities. Thus, the model is extended to control for the differences in each municipality's steady state output levels and include the spatial effects as:

$$\frac{1}{T}log\left[\frac{y_{i,t_0+T}}{y_{i,t_0}}\right] = (pWY) + a + \beta_0 log\left[y_{i,t_0}\right]\sum_{i=1}^{k}\beta_k X_{i,t_0} + u_{i,t_0+t_0+T} \tag{2}$$

This extended model includes the initial level of per-capita production, control variables that reflect the differences in each municipality's steady state output levels, and a continuity matrix (*W*) to capture the geographical context. In Eq. 2, $\beta_0$ measures the convergence speed towards the steady state per-capita production level. If convergence exists, $\beta_0$ will be negative while a positive sign of $\beta_0$ will indicate divergence in per-capita levels.

The control variables are: i) people with elementary school per 100,000 inhabitants; ii) natural logarithm of gross capital formation; iii) natural logarithm of the interaction between initial production level and gross capital formation; iv) percentage of big firms within each municipality; and v) logarithm of the interaction between initial production level and people with elementary school per 100,000 inhabitants. Equation 2 is estimated using maximum likelihood with spatially lagged effects (spatial lag regression) and Geographically Weighted Regression (GWR).

The first step of the analysis consists in determining whether our dependent variable is a stochastic phenomenon or contrary. The statistic Moran's I is widely employed for testing the presence of spatial dependence in observations. This provides a global statistic for assessing the degree of spatial autocorrelation between observations as a function of the distance separating them.

The device typically used in spatial analysis is the so-called spatial weight matrix, or simply W matrix. Impose a structure in terms of what are the neighbors for each location. Assigns weights that measure the intensity of the relationship among pair of spatial units.

The spatial weight matrix, W, a *nxn* positive symmetric and non-stochastic matrix with element $w_{ij}$ at location *i, j*. The values of $w_{ij}$ or the weights for each pair of locations are assigned by some preset rules which defines the spatial relations among locations. By convention, $wij = 0$ for the diagonal elements. Two main approaches: contiguity or based on distance.

$$\mathrm{W} = \begin{pmatrix} w_{11} & w_{12} & \dots & w_{1n} \\ w_{21} & w_{22} & \dots & w_{2n} \\ \vdots & \vdots & \ddots & \vdots \\ w_{n1} & w_{n1} & \dots & w_{nn} \end{pmatrix} \tag{3}$$

The availability of polygon or lattice data permits the construction of contiguity-based spatial weight matrices. A typical specification of the contiguity relationship in the spatial weight matrix is

$$w_{ij} = \begin{cases} 1 & \text{if } i \text{ and } j \text{ are contiguous} \\ 0 & \text{if } i \text{ and } j \text{ are not contiguous} \end{cases} \tag{4}$$

Binary Contiguity: Rook's case (Common Border), Bishop's case (Common Vertex), Queen's (Kings) case (Either common border or vertex). See Fig. 1.

Weights may be also defined as a function of the distance between region i and j, $d_{ij}$. $d_{ij}$ is usually computed as the distance between their centroids (another important unit). Let $x_i$ an $x_j$ be the longitude and $y_i$ and $y_j$ the latitude coordinates for region i and j, respectively.

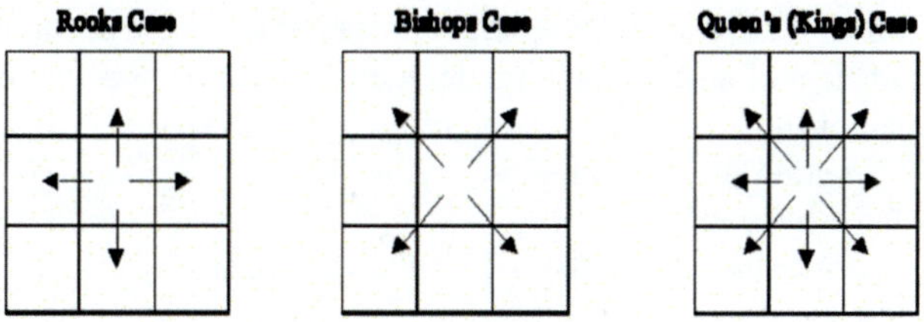


**Fig. 1.** Binary contiguity.

The local Moran's I or as known, local indicator of spatial association (LISA), it allows the decomposition of the global indicator which in turn helps to explore the extent of significant clustering with values similar in magnitude around a particular observation [8].

The Global Moral Index equation is given by

$$I = \left(\frac{N}{\sum_i \sum_j w_{ij}}\right)\left(\frac{\sum_i \sum_j w_{ij}(x_i - \overline{x})(x_j - \overline{x})}{\sum (x_i - \overline{x})^2}\right) \tag{5}$$

Where *N* is the number of spatial units denoted by *i* and *j*, *X* is the mean, *x* is the variable of interest and *wij* is element *ij* of the spatial weighting matrix. And the local Moran's I o LISA equation is given by

$$I_i = z_i \sum_j w_{ij} z_j \; Z_i = \frac{x_i \overline{x}}{SD_x} \tag{6}$$

Where $z_i$ and $z_j$ are the standard deviations of the mean of the x's (standardized).

In search of local variations of spatial autocorrelation, or spatial clusters, we perform the respective LISA statistic. In doing this, we can identify municipalities with concentration of high values, concentration of low values, and spatial outliers. The maps show the distribution of four different types of spatial clustering: a) High-High (HH): a municipality with high value and its neighbors too; b) High-Low (HL): a municipality with a high outlier and its neighbor's low values; c) Low-High (LH): a municipality with a low outlier and its neighbors high values; d) Low-Low (LL): a municipality with a low value and its neighbors too.

Our next step is to estimate spatial econometric models in order to account for spatial dependence. The spatial weight matrices (W) represent the "degree of potential interaction" between neighboring locations [9]. The parameters ρ is scalar spatial parameters measuring the degree and type of spatial dependence. For example, suppose the case in which $\rho \neq 0$, the resulting is what is called a spatial lag model or spatial autoregressive model. An appropriate estimation technique is Maximum Likelihood Estimation (MLE) which gives consistent and efficient parameters [9].

Geographical Weighted Regression (GWR) is another method that allows regression coefficients systematically vary over space. With this approach, separate regression models are estimated for each areal unit. The farer away a spatial unit from the region for which the regression is fitted, the lower its data is weighted.

The GWR method accounts for spatial nonstationary in the economic relationship. Response parameters of an economic model may be location specific. From a theoretic point of the relationships are intrinsically different across space.

The GWR equation for the *ith* region is given by

$$y_i = \beta_1(u_i, v_i) + \beta_2(u_i, v_i) \cdot x_{i2} + \ldots + \beta_k(u_i, v_i) \cdot x_{ik} + \varepsilon_i \tag{7}$$

The location specific regression coefficients $\beta_j(u_i,v_i)$ are functions of longitude and latitude coordinates $u_i$ and $v_i$. More precisely, $\beta_j(u_i,v_i)$ is a realization of the continuous function $\beta_j(u_i,v_i)$ at point i.

The local parameters $\beta_j(u_i,v_i)$ are estimated by a weighted least squares procedure. The weights $w_{ij}$, $j = 1, 2,\ldots, n$, at each location $(u_i, v_i)$ are defined by a function of the distance $d_{ij}$ between the center of the region i and those of the other regions. We have to estimate n unknown vectors of local regression coefficients:

$$\beta(\mathbf{1}) = \begin{bmatrix} \beta_1(u_1, v_1) \\ \beta_2(u_1, v_1) \\ \vdots \\ \beta_k(u_1, v_1) \end{bmatrix}, \beta_\mathbf{2} = \begin{bmatrix} \beta_1(u_2, v_2) \\ \beta_2(u_2, v_2) \\ \vdots \\ \beta_k(u_2, v_2) \end{bmatrix}, \cdots, \beta(\mathbf{n}) = \begin{bmatrix} \beta_1(u_n, v_n) \\ \beta_2(u_n, v_n) \\ \vdots \\ \beta_k(u_n, v_n) \end{bmatrix}, \tag{8}$$

The GWR estimates of the unknown local parameter vector $\beta(i)$ are given by

$$\hat{\boldsymbol{\beta}}(\boldsymbol{i}) = [\boldsymbol{X} \cdot \boldsymbol{W}(\boldsymbol{i}) \cdot \boldsymbol{X}]^{-1} \cdot \boldsymbol{X} \cdot \boldsymbol{W}(\boldsymbol{i}) \cdot \boldsymbol{y},\ \boldsymbol{i} = \mathbf{1}, \mathbf{2}, \ldots, \boldsymbol{n} \tag{9}$$

**W(i)** is the *nxn* spatial weight matrix which has the form

$$\boldsymbol{W}(\boldsymbol{i}) = \begin{bmatrix} w_{i1} & \mathbf{0} & \cdots & \mathbf{0} \\ \mathbf{0} & w_{i2} & \cdots & \mathbf{0} \\ \vdots & \vdots & \ddots & \vdots \\ \mathbf{0} & \mathbf{0} & \cdots & w_{in} \end{bmatrix} \tag{10}$$

## 5 Data

The information for this study comes from the Economic Census, which is collected by the National Institute of Statistics and Geography of Mexico (INEGI) every five years. Information regarding education is estimated based on the Population Census and the Projections of Population conducted by Mexico's National Population Council (CONAPO). The infrastructure indicators and municipal government expenditures are obtained from INEGI. The years of study are 2013 and 2018. We use deflators from the Central Bank of Mexico (BANXICO) to correct inflation problems. The dependent variable *y* is calculated by subtracting the natural log of production of firms in municipality *i* of the previous period from the natural log of GDP of the current period and multiplying by one-fifth.

## 6 Results

Figure 2 and Fig. 3 presents the production level of firms by city in 2013 and 2018, respectively. Darker colors indicate higher levels of production. As it can be seen, higher levels of production of firms are mainly located in the northern border of Mexico and there is a high concentration of production value in the center of Mexico. In contrast, southern cities have lower production of firms both in 2013 and 2018.

**Fig. 2.** Production level of firms by city in 2013. Source: Own elaboration with information form the Economic Census, 2014.

**Fig. 3.** Production level of firms by city in 2018. Source: Own elaboration with information form the Economic Census, 2019.

In order to determine the type of continuity matrix to include in Eq. 2, the Moran's I for the dependent variable is estimated for seven forms of continuity matrix as shown in Table 1. Although Table 1 shows the Moran's I for the production level of firms in

2013 and 2018, the interest is in the logarithm of the ratio of both years, which is the dependent variable in Eq. 2. The continuity matrix type Rook 1 has the highest value of Moran's I. Therefore, we use this type of matrix for Eq. 2.

**Table 1.** Moran's I for different continuity matrices

| Continuity Matrix | Production level of firms in 2013 (y13) | Production level of firms in 2018 (y18) | log(y18/y13) |
|---|---|---|---|
| Queen_1 | 0.1075 | 0.0589 | 0.0522 |
| Queen_2 | 0.0851 | 0.0718 | 0.0339 |
| Queen_3 | 0.0523 | 0.0348 | 0.0145 |
| Rook_1 | 0.1095 | 0.0599 | 0.0539 |
| Rook_2 | 0.0857 | 0.0716 | 0.0328 |
| Rook_3 | 0.0523 | 0.0351 | 0.0166 |
| Distance | 0.0250 | 0.0156 | 0.0122 |

Source: Own estimation

Figure 4, Fig. 5 and Fig. 6. show the spatial correlation of production level of firms in 2013, 2018, and the logarithm of production of firms in 2018 divided by production level in 2013, respectively. As it can be seen in Fig. 3 and Fig. 4, there is a high correlation of cities with high level of production of firms in Monterrey-Saltillo region and Chihuahua northern border both in 2013 and 2018. Meanwhile, cities in Oaxaca, Guerrero and

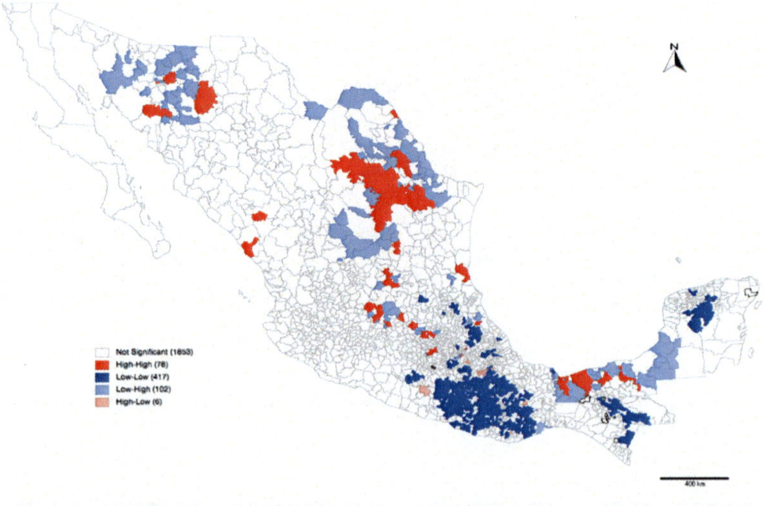


**Fig. 4.** Spatial correlation of production of firms in 2013. Source: Own elaboration with information form the Economic Census, 2014.

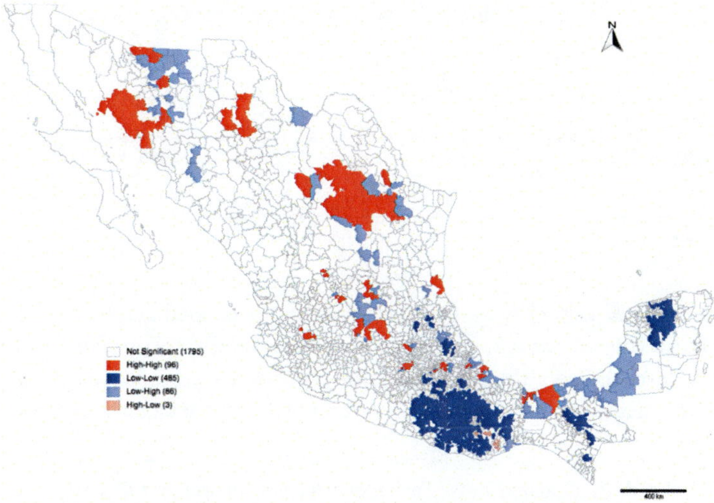


**Fig. 5.** Spatial correlation of production of firms in 2018. Source: Own elaboration with information form the Economic Census, 2019.

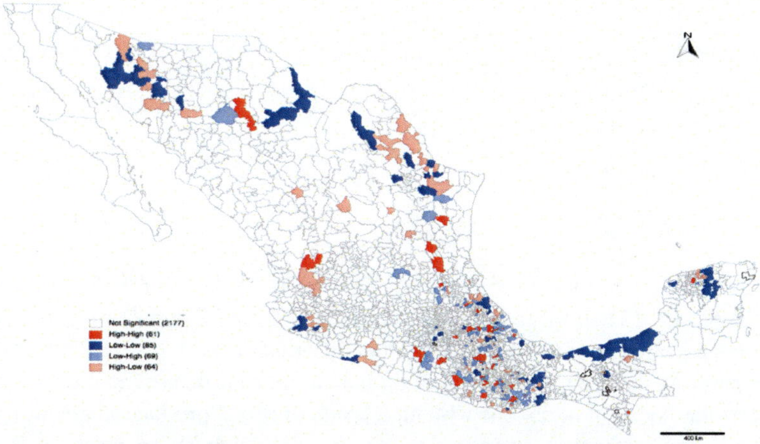


**Fig. 6.** Spatial correlation of dependent variable. Source: Own elaboration with information form the Economic Census, 2009 and 2014.

Chiapas state concentrate a high correlation of firms with low production in 2013 and 2018.

Our procedure for calculating our additional results was compiled through a series of steps that began with organizing and distributing our data. After determining the continuity matrix with the highest Moran's I, we tested the normality of errors. The Jarque-Bera test indicates that there is no a multicollinearity problem in the data. Table 2 shows three tests to determine the existence of heteroskedasticity. The three indicators show the existence of heteroskedasticity, but it can be attributed to the fact that the error variance has a spatial dependence.

**Table 2.** Tests for heteroskedasticity

| TEST | DF | VALUE | PROB |
|---|---|---|---|
| Breusch-Pagan test | 6 | 102.71 | 0.0001 |
| Koenker-Bassett test | 6 | 22.29 | 0.0011 |
| White | 27 | 336.29 | 0.0001 |

Source: Own estimation

Table 3 shows a series of diagnostics for spatial dependence for weight matrix Rook 1. Based on results of Table 3, there is evidence of spatial autocorrelation. The Lagrange Multiplier tests indicate that both types of regressions are adequate (spatial lag and spatial error). However, the spatial lag regression is more adequate for data.

**Table 3.** Diagnostics for spatial dependence for weight matrix Rook 1

| TEST | MI/DF | VALUE | PROB |
|---|---|---|---|
| Moran's I (error) | 0.048 | 4.011 | 0.0001 |
| Lagrange Multiplier (lag) | 1 | 17.564 | 0.0001 |
| Robust LM (lag) | 1 | 7.586 | 0.0059 |
| Lagrange Multiplier (error) | 1 | 15.402 | 0.0001 |
| Robust LM (error) | 1 | 5.424 | 0.0199 |
| Lagrange Multiplier (SARMA) | 2 | 22.986 | 0.0001 |

Source: Own estimation

Results of the spatial lag regression are shown in Table 4. According to the results, from 2008 to 2013 the positive sign of the coefficient for the initial production indicates a divergence process of firms' production in different cities in Mexico. Results support the hypothesis that geographical location is a significant determinant of growth. This suggests that location in regions with high levels of firms' production positively affects the growth of production, and firms located in cities with higher levels of production reported greater growth than firms in municipalities with lower firms' production levels.

The coefficients related to the number of big firms and education improvements amongst the population are not significant. However, the interaction between schooling and initial production level is negative and significant. This suggests that, holding the initial per capita production level constant, cities with lower firms' production and higher levels of human capital exhibited faster convergence rates. On the other hand, while the coefficient of gross capital formation is not significant, its interaction with firms' production level is negative. This negative sign indicates that the production level of cities with lowers firms' production and higher levels of gross capital formation will increase more than the production of municipalities with higher production and gross capital formation. The number of big firms is not a determinant of growth, but the strategic location of firms.

**Table 4.** Results of the spatial lag regression

| Variable | Coefficient | z | Prob |
|---|---|---|---|
| Continuity matrix | 0.126554 | 4.0685 | 0.0001* |
| Constant | 0.019547 | 2.0558 | 0.0398* |
| Log initial production (Y13) | 0.000550 | 5.5287 | 0.0001* |
| Log gross capital formation 2013 (GCF13) | 4.432e012 | 1.8460 | 0.0649 |
| Log(Y13* GCF13) | 2.512e009 | 3.2496 | 0.0012* |
| Education 2013 (EDU13) | 0.033505 | 1.8312 | 0.0670 |
| Log(Y13*EDU13) | 0.001429 | 5.9613 | 0.0001* |
| Big firms 2013 | 5.581e010 | 1.0982 | 0.2721 |

$R^2$: 0.4215 *significant at 0.05.
Source: Own estimation.

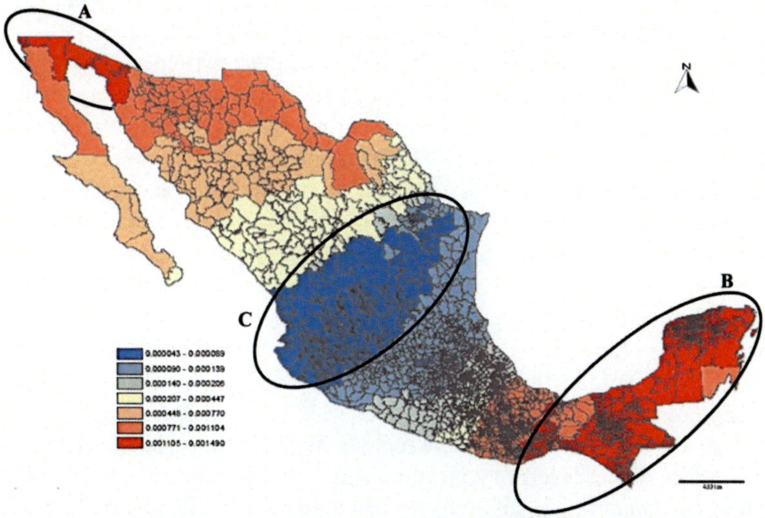


**Fig. 7.** Convergence Estimator Using Weighted Geographic Regression. Source: Own elabora tion with information form the Economic Census, 2009 and 2014.

Figure 7 shows the estimates of convergence using GWR, which indicates that the level of the divergence process of firms' production is heterogeneous throughout the country. Regions A and B are those where there is a greater level of divergence process of firms' production. These areas are the south-southeast of Mexico and the northwestern border of Mexico with the United States. The north central region (C) is where there is the least divergence process of firms' production.

## 7 Conclusions

From 2013 to 2018 the positive sign of the coefficient for the initial production indicates a divergence process of firms' production in different municipalities in Mexico. Results support the hypothesis that geographical location is a significant determinant of growth. This suggests that location in regions with high levels of firms' production positively affects the growth of production, and firms located in municipalities with higher levels of production reported greater growth than firms in municipalities with lower firms' production levels.

With several different reforms being created since the late 1980s, expanding economic potential of Mexican regions has become an opportunity in which many firms aim to achieve. Not only would the benefit of a well reformed Mexican economic infrastructure be beneficial to the firms based in the country, but the additional trade routes, innovative production opportunities, and employment possibilities would increase the overall GDP performance of Mexico, which would consequently positively increase the overall global economy. Through reforms such as NAFTA and GATT (the General Agreement of Tariffs and Trade), the economic potential and growth patterns projected for Mexico's economy were targeted to correct regional growth issues.

Although initial conditions, such as cities with better equipment, higher capabilities of industrial and human capital, along with better transportation and communication avenues being readily available in regions during reforms, are important determinants of production growth, the results of this study show that schooling and gross capital formation may reduce inequalities in production level of firms in different regions. This indicates that reforms to reduce differences in regions should be geared toward reducing inequalities in educations between poorer and richer regions in Mexico.

Northern Mexico's modernization and improved education, especially in manufacturing firms, has dramatically increased and improved with the United States. As it stands, it is more expensive and difficult to establish a manufacturing firm close to the U.S. border where productivity and efficiency is at its highest. Therefore, regions farther from the border cannot produce products as well as the northern region, and struggle to innovate as successfully as northern regions. With less opportunities for the southern regions to gain higher levels of education, and economic growth; the overall GDP and growth of the country have suffered. As stated previously, Mexico is notorious for the high levels of inequality among its different regions throughout the country. This trend shows no inclination of changing.

In order to promote substantial economic growth throughout Mexico and increase the life of firms, the country must create avenues for the prosperity and success of its northern regions to trickle down evenly into the south. One way for this could be achieved is through aggressive focus on developing cities closer to the southern border as a whole. For example, making more southern region Mexican cities more ideal tourist locations similar to Cancun, would generate wealth in areas that could convert the excess cash into manufacturing facilities. If more southern region cities are modernized to resemble the already prominent northern cities infrastructure, then firms will be more willing and confident to invest in growing the economy.

Also, results of this study show that location determines the successful of firms. This indicates that firms' production increases based on cluster. According to Porter

[5] cluster helps firms' competitiveness in three ways: i) increases firms' productivity; ii) improves capacity of innovation; and iii) promotes the launch of new firms that support innovation. In order to increase firms' regional growth, economic policies should promote the formation of cluster and integrate small businesses into productive chains.

## References

1. INEGI. Sistema de Cuentas Nacionales. Aguascalientes: Instituto Nacional de Geografía y Estadística (2020)
2. Arroyo, F.: Dinámica del PIB de las entidades federativas de México, 1980–1999. Comercio Exterior **51**(7), 583–600 (2001)
3. Cermeño, R.: Decrecimiento y Convergencia de los Estados Mexicanos. Un Análisis de Panel. El Trimestre Económico **68**(4), 603–629 (2001)
4. Chiquiar, D.: Why Mexico's regional income convergence broke down. J. Dev. Econ. **77**(1), 257–275 (2005)
5. Porter, M.: On Competition. Harvard Business Press, Boston (1998)
6. López, J.: Teorías y Enfoques del Desarrollo Regional. Bógota: Escuela Superior de Administración Pública (2003)
7. Barro, R., Sala-i-Martin, X.: convergence. J. Poli. Eco. **100**(2), 223–251 (1992)
8. Lloyd, C.: Local Models for Spatial Analysis, 2nd edn. CRC Press, Taylor & Francis Group, Boca Raton, Florida (2007)
9. Anselin, L.: Spatial Econometrics: Methods and Models. Kluwer Academic Publishers, Dordecht (1988)
10. INEGI. Economic Census 2014 and 2019. Aguascalientes: Instituto Nacional de Geografía y Estadística (2021)

# On-Chain Metrics and Technical Analysis in Cryptocurrency Markets

Angel Roberto Nava-Solis and Eduardo Javier Treviño-Saldívar(✉)

Facultad de Contaduría Pública y Administración, Universidad Autónoma de Nuevo León, Av. Universidad S/N, Cd. Universitaria, 66451 San Nicolás de los Garza, NL, Mexico
{angel.navas,eduardo.trevinosl}@uanl.edu.mx

**Abstract.** This study attempts to analyze comprehensive data for the cryptocurrency market. In detail, this paper gathers all pertinent market currency data, on-chain data, and technical analysis to predict investor sentiment. In this research, we analyze the most effective tools to predict the market price of cryptocurrencies, from 2016 to 2021 and to determine the market price trend and its forecasts, through complete historical data in on-chain and crypto-currency market price. The methodology and tools used to analyze this crypto assets, provide a more effective understanding, through on-chain metrics indicators, with increase results in the accuracy on forecasting the crypto-currency prices in the future. Back tests are examined as evidence of the reliability of these indicators such as cryptographic portfolio sizes, sales, purchases and increase of relevant economic activity of the blockchain network to anticipate the crypto market prices rises.

**Keywords:** Blockchain · Cryptocurrencies · Investor sentiment

## 1 Introduction

Blockchain started thirteen years ago in a globalized world, when Satoshi Nakamoto build the model and its concept in 2008; even though it was implemented a year later, through Bitcoin, a digital payment system and cryptocurrency, creating the first cryptocurrency, as a virtual assets and innovating the currency financial market [1].

The cryptocurrency market has been a trending topic ever since the past decade, gathering technological power, attracting attention and investments worth trillions of dollars on a global scale. Cryptocurrency technology and its network has had many superior features due to its unique architecture, which also, determined its efficiency, applicability, and data-intensive characteristics around the world.

This article shows the latest cryptocurrency market price prediction tools, the differences and characteristics will be discussed. The objective is to determine which tools are the most effective under certain conditions. For the cryptocurrency market, on-chain metrics will be thoroughly analyzed. In a conservative point of view, technical analysis, Exponential Moving Averages (EMA) and the Moving Average Convergence Divergence (MACD) tools will be used to verify the efficacy of this methods. These tools will be used to predict the price of crypto assets in general during the period 2017 to 2021.


M. d. P. Rodríguez García et al. (Eds.): XX SIGEF 2021, LNNS 384, pp. 122–129, 2022.
https://doi.org/10.1007/978-3-030-94485-8_10

## 2 Literature Review

Since 2017, the price of the first cryptocurrencies such as Bitcoin, Ethereum, Litecoin Ripple began to trade on the Chicago Mercantile Exchange (CME), which contributed to the reversal of the bitcoin price analyzed by Hale *et al.* [2]. Before December 2017, there was no market for bitcoin derivatives.

Yermack [3] mentions that the price of bitcoin cannot be sold short, which is why the value of bitcoin grew over time due to the absence of financial derivatives or options on bitcoin futures that correct the crypto currency market, so the price was determined due to the sale of the asset in term of points, in the cryptocurrency exchange and by not adding short options in the futures contracts. This analysis was based on metrics on derivatives on-chain, futures open interest, volume of long futures settlements, and short futures settlements.

As market conditions and asset trends changed, the need to study these metrics is essential to be more efficient in creating value in crypto currency market trading. Since beginning, bitcoin price volatility is very high. Peterson [4] mentions that with 100% of confidence that the price of Bitcoin has been fraudulently manipulated at some point in its lifespan since 2010 and with 98% of confidence that bitcoin was manipulated until 2019 and any traditional technical analysis method was not effective in predicting its trend. Hayes [5] evaluates a method to predict the equilibrium price, expressed in dollars per bitcoin, in which the marginal cost of production is equal to the marginal profit, that is,

$$P = \mathrm{Eday}\frac{\mathrm{BTC}}{\mathrm{day}} = \$\,\mathrm{kWh}\cdot\mathrm{EEF}\cdot\mathrm{hrday}\cdot\delta\beta\cdot 1000\frac{\mathrm{GH}}{\mathrm{s}}\cdot\theta \tag{1}$$

where Eday is the cost per day for a producer, \$/kWh is the price of a kilowatt-hour and EEF is the energy consumption efficiency of the miner's hardware that are the basics of the on-chain analysis.

## 3 Theoretical Frameworks

On-chain analysis is the most important disruptive trend in the crypto asset investment industry, as it uses metrics from the blockchain of the underlying asset. Crypto assets represent a significant innovation over traditional forms of financing and investing, but conventional financial valuation metrics are insufficient, incomplete, and ineffective to analyze these new types of assets.

Zhao *et al.* [6] mentions that blockchain metrics include terminology as miners' income, extraction rate, gas, active addresses, hash rate, and the difficult level on analyzing the crypto asset, and all these variables are a basis for crypto market price prediction. Meanwhile, blockchains constantly generate a large amount of open and incorruptible financial data that allows us to access accurate and reliable measures of relevant economic activity on crypto networks. By analyzing this on-chain data, we can change the way we measure market sentiment and behavior. Hassani *et al.* [7] talk about the significant impact of cryptocurrency being inextricably linked to blockchain technology, which is the key to Bitcoin and the fundamental technique for other cryptocurrencies.

Cryptocurrency market can be examined with the same technical analysis indicators that are seen in Forex, stocks, and commodities trading, such as, relative strength index (RSI), exponential moving average (EMA), moving average (MA), Cross, moving average convergence divergence (MACD) and Bollinger bands (BA) that are used to predict the behavior of the market prices. Furthermore, crypto technical analysis (On-chain) metrics are also analyzed, that is, massive data sets of investor activity through the public ledger of each cryptographic asset, in which the intrinsic value can be analyzed using tools such as hash rate analytics, daily active addresses, balance on exchanges, difficulty ribbon and number of addresses with balance $\geq$0.01 that analyzes the metrics for each-chain action in history (Jagannath *et al.*) [8].

Greaves *et al.* [9] states from the bitcoin blockchain contains various network-based functions as important variables such as flow characteristics and measurements of node mining. Also, it is important to clarify that on chain metrics is limited in including certain exchange information from bitcoin exchanges.

### 3.1 Main On-Chain Signals. Bitcoin: Mean Hash Rate

The "hash rate" is the unit of measurement of the processing power of the Bitcoin network. The Bitcoin network must do intensive math operations for security reasons. When the network reaches a hash rate of 10 TH/s it means that it can do 10 trillion calculations per second. Hayes [5] demonstrated the difference between the computational power used by a miner and its expected benefit given current network conditions (Fig. 1).

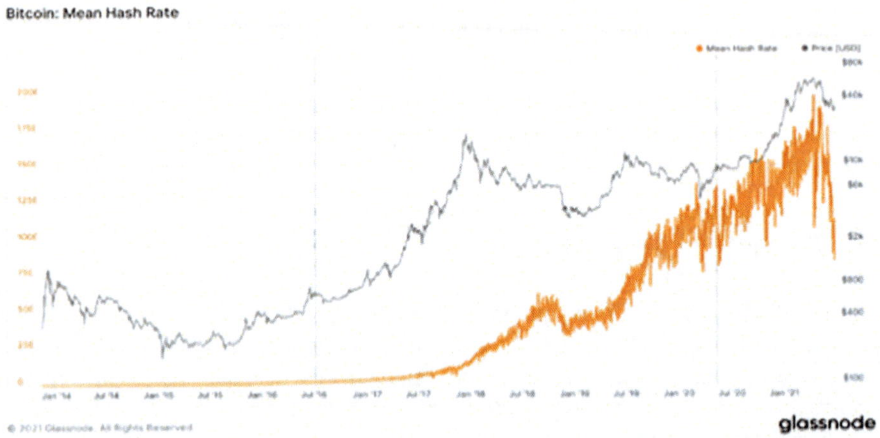


Fig. 1. Bitcoin: mean hash rate. Source: graphics provided by Glassnode software.

### 3.2 Daily Active Addresses

Zhao, D. et al. [6] stablished the number of active addresses on the network and its numbers of transactions sent and received on the network. Successful active addresses generally denote a demand side feature while the native coin emission rate is the supply side. (See Fig. 2).

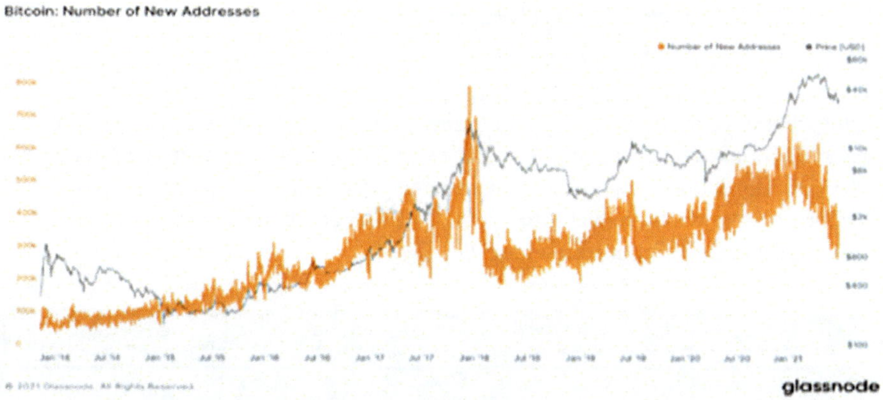


**Fig. 2.** Bitcoin: number of new addresses. Source: graphics provided by Glassnode software.

### 3.3 "GBTC". Grayscale Flows

Trust Mutual Fund owns 654,885 Bitcoin, or 46% of the 1.4 million Bitcoin held by publicly traded companies, according to Bitcointreasuries.org. As of April 8, 2021, GBTC is currently trading at $ 47.57 and, according to official documents, it has 0.00095 Bitcoin (with a value of $ 54.6) per share, this indicator can be the results of movements from purchases or sales of the institutional investment fund and can be used as an indicator for buying and selling analysis tool of investment fund grayscale (See Fig. 3).

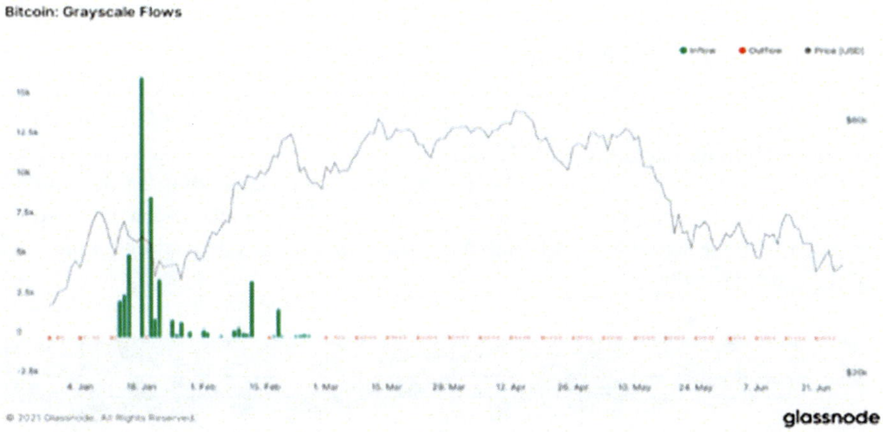


**Fig. 3.** Bitcoin: grayscale. Source: graphics provided by Glassnode software.

## 3.4 Technical Analysis in Cryptocurrencies

### EMA – Exponential Moving Average

The Exponential Moving Average or EMA, is one of the basic technical analysis indicators that is very useful for cryptocurrency traders to determine the price trend of an asset. EMA takes into account the average price of the asset of a certain period of time. The indicator draws a line of "n" day average price in the line graph chart, measuring the long time trend through crossing with the EMA graph line signaling bear market or bull market in the short or long-term. (See Fig. 4).

**Fig. 4.** Bitcoin: Ema. Source: graphics provided by tradingview software.

### MACD – Moving Average Convergence Divergence

The MACD indicator provides statistical information about the price trend of an assets in a determined market. This tool analyzes the level at which the lines and price levels are in a bull trend market: the MACD histogram appears above the zero line, together with the signal line Y will mark a blue point on the cross depending on the timing of the chart. In a bear market: the MACD histogram will be located below the zero line, together with the signal line will mark a red point on the cross depending on the timing of the chart. (See Fig. 5).

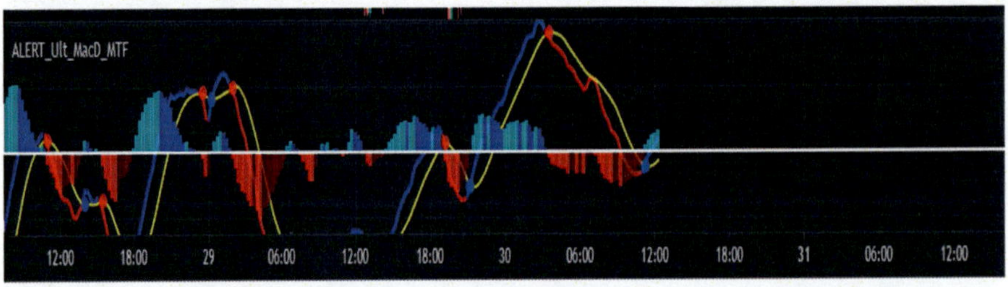


**Fig. 5.** Bitcoin: MACD. Source: graphics provided by tradingview software.

### 3.5 Hypothesis

Previous studies have explored different technical analysis tools to analyze prices of financial assets such as the price of crypto assets, but there is no evidence to have an exhaustive research that compares this tools along with "on-chain" analysis in crypto assets Therefore, the following hypothesis will be:

> $H_1$: ON chain metric tool is more effective in long term price forecasting in crypto currency assets, in order to have a better prediction tool in the investing decision process.

## 4 Data and Method

### 4.1 Data

A new technical analysis strategy proposal is made in crypto assets based on on-chain metrics. To execute the analysis, we use the CME futures price "BTC1!", which shows the spot price of the main asset cryptographic chosen (bitcoin) and analyzes its behavior with different types of on-chain analysis. The timespan considered is from the year 2016 to 2021.

### 4.2 Method

We use the hash ribbons halving script on the trading view platform as a learning algorithm to analyze price prediction (See Fig. 6). For the first technical analysis strategy, it will be highlighted in the trading view platform to analyze the price of bitcoin with the most popular indicators of traditional technical analysis such as: EMA, MACD and RSI in a daily timeframe, and for the technical analysis using on metrics-chain the glassnode platform will be used on the bitcoin price with metrics: mean hash rate, active addresses, balance on exchanges, difficulty ribbon and Number of Addresses with Balance $\geq$0.01.

A we carried out the analysis from the year 2016 to 2021, the script added the halving 2016 and 2021 together with the data of the hast rate, giving data of the beginning of the capitulation of mining red interpreted as sale or bear market, new capitulation (spring) blue interpreted as buying or starting bull market.

## 5 Results

The price prediction from 2016 to 2021 was made using the on-chain hash ribbon halving tools and in technical analysis, EMA and MACD. As you can see in Fig. 7, the metrics and on-chain analysis can be better anticipate the buy and sell "signal" according to the capitulation of the mining, halving and decrease of the hash rate. Table 1 shows the results of halving onset, capitulation onset and spring.

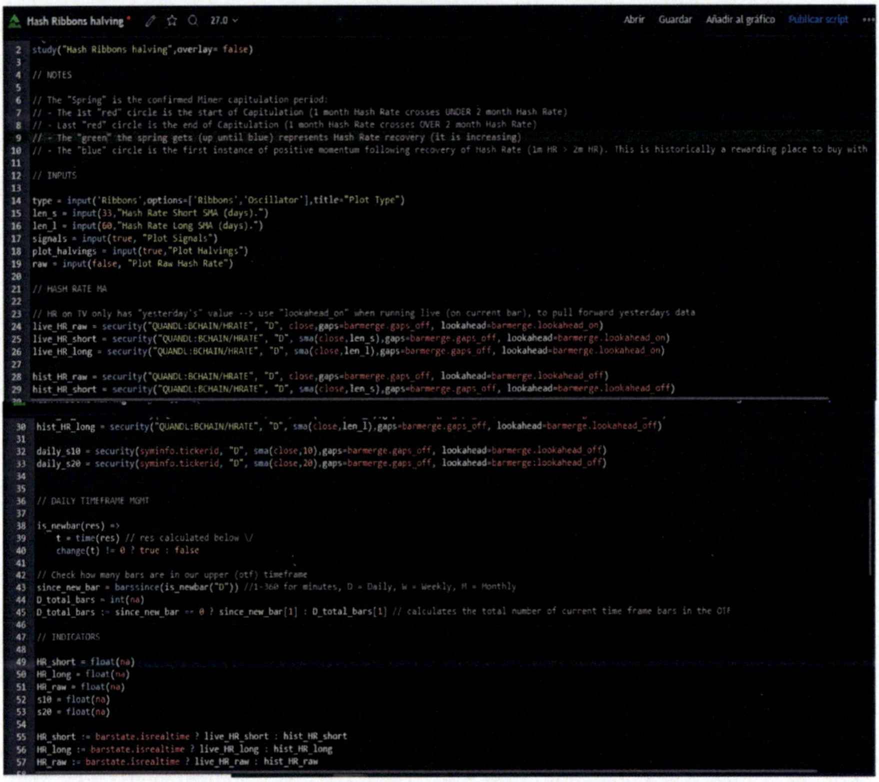


```
study("Hash Ribbons halving",overlay= false)

// NOTES

// The "Spring" is the confirmed Miner capitulation period:
// - The 1st "red" circle is the start of Capitulation (1 month Hash Rate crosses UNDER 2 month Hash Rate)
// - Last "red" circle is the end of Capitulation (1 month Hash Rate crosses OVER 2 month Hash Rate)
// - The "green" the spring gets (up until blue) represents Hash Rate recovery (it is increasing)
// - The "blue" circle is the first instance of positive momentum following recovery of Hash Rate (1m HR > 2m HR). This is historically a rewarding place to buy with

// INPUTS

type = input('Ribbons',options=['Ribbons','Oscillator'],title="Plot Type")
len_s = input(33,"Hash Rate Short SMA (days).")
len_l = input(60,"Hash Rate Long SMA (days).")
signals = input(true, "Plot Signals")
plot_halvings = input(true,"Plot Halvings")
raw = input(false, "Plot Raw Hash Rate")

// HASH RATE MA

// HR on TV only has "yesterday's" value --> use "lookahead_on" when running live (on current bar), to pull forward yesterdays data
live_HR_raw = security("QUANDL:BCHAIN/HRATE", "D", close,gaps=barmerge.gaps_off, lookahead=barmerge.lookahead_on)
live_HR_short = security("QUANDL:BCHAIN/HRATE", "D", sma(close,len_s),gaps=barmerge.gaps_off, lookahead=barmerge.lookahead_on)
live_HR_long = security("QUANDL:BCHAIN/HRATE", "D", sma(close,len_l),gaps=barmerge.gaps_off, lookahead=barmerge.lookahead_on)

hist_HR_raw = security("QUANDL:BCHAIN/HRATE", "D", close,gaps=barmerge.gaps_off, lookahead=barmerge.lookahead_off)
hist_HR_short = security("QUANDL:BCHAIN/HRATE", "D", sma(close,len_s),gaps=barmerge.gaps_off, lookahead=barmerge.lookahead_off)
hist_HR_long = security("QUANDL:BCHAIN/HRATE", "D", sma(close,len_l),gaps=barmerge.gaps_off, lookahead=barmerge.lookahead_off)

daily_s10 = security(syminfo.tickerid, "D", sma(close,10),gaps=barmerge.gaps_off, lookahead=barmerge.lookahead_off)
daily_s20 = security(syminfo.tickerid, "D", sma(close,20),gaps=barmerge.gaps_off, lookahead=barmerge:lookahead_off)


// DAILY TIMEFRAME MGMT

is_newbar(res) =>
    t = time(res) // res calculated below \/
    change(t) != 0 ? true : false

// Check how many bars are in our upper (otf) timeframe
since_new_bar = barssince(is_newbar("D")) //1-360 for minutes, D = Daily, W = Weekly, M = Monthly
D_total_bars = int(na)
D_total_bars := since_new_bar == 0 ? since_new_bar[1] : D_total_bars[1] // calculates the total number of current time frame bars in the OTF

// INDICATORS

HR_short = float(na)
HR_long = float(na)
HR_raw = float(na)
s10 = float(na)
s20 = float(na)

HR_short := barstate.isrealtime ? live_HR_short : hist_HR_short
HR_long := barstate.isrealtime ? live_HR_long : hist_HR_long
HR_raw := barstate.isrealtime ? live_HR_raw : hist_HR_raw
```

**Fig. 6.** Script hash ribbons. Source: code provided by tradingview software.

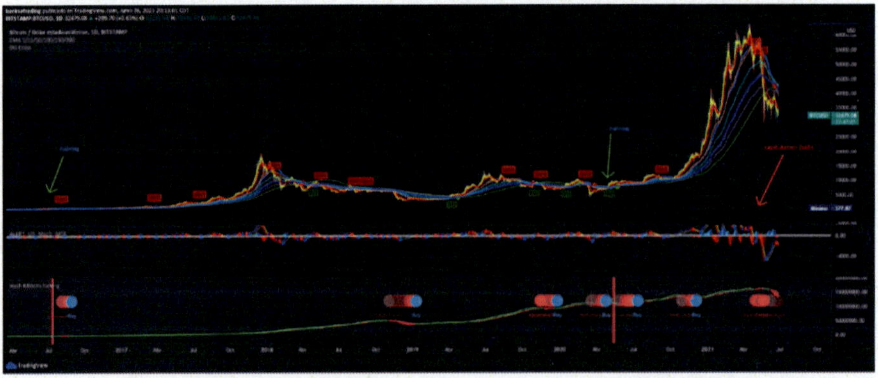

**Fig. 7.** Graph motion with script hash ribbons. Source: graphics provided by tradingview software.

**Table 1.** Predictive analysis of script hash ribbons for the Bitcoin

| Year | Bitcoin price (spring) | Capitulation of the mining | Halving date | Performance |
|---|---|---|---|---|
| 2016–2017 | $ 767 usd | 9,679 usd | 11-06-2016 | $ 8,912 usd |
| 2020–2021 | 3,9818 usd | 46,509 usd | 11-05-2020 | $ 42,591 usd |

## 6 Conclusions

As we can see in the findings of this research, On Chain metrics is more effective in forecasting price prediction in crypto markets, market cycles, their faces, trends were identified in advance of the technical analysis, it was detected that it was the best option for trading, using hash ribbons (on chain), since it provides greater confidence, due to the data obtained from the blockchain, which is planned to be invested, as the most rational option. Quantitative analysis metrics (chain) is a better tool for decision making in long-term investment strategies, due to include a broader vision of the cryptographic market.

## References


1. Hyland-Wood, D., Khatchadourian, S.: A future history of international blockchain standards. J. Br. Blockchain Assoc. **1**(1), 25–31 (2018)
2. Hale, G., Krishnamurthy, A., Kudlyak, M., Shultz, P.: How futures trading changed bitcoin prices. FRBSF Econ. Lett. **12**, 1–5 (2018)
3. Yermack, D.: Is Bitcoin a real currency? An economic appraisal (No. w19747). Natl. Bureau Econ. Res. **36**(2), 843–850 (2013)
4. Peterson, T.: To the Moon: A History of Bitcoin Price Manipulation. Available at SSRN 3639431 (2020)
5. Hayes, A.S.: Cryptocurrency value formation: an empirical study leading to a cost of production model for valuing bitcoin. Telematics Inform. **34**(7), 1308–1321 (2017)
6. Zhao, D., Rinaldo, A., Brookins, C.: Cryptocurrency price prediction and trading strategies using support vector machines. arXiv Preprint arXiv:1911.11819 (2019)
7. Hassani, H., Huang, X., Silva, E.: Big-crypto: big data, blockchain and cryptocurrency. Big Data Cognit. Comput. **2**(4), 34 (2018)
8. Jagannath, N., et al.: A self-adaptive deep learning-based algorithm for predictive analysis of bitcoin price. IEEE Access **9**, 34054–34066 (2021)
9. Greaves, A., Au, B.: Using the bitcoin transaction graph to predict the price of bitcoin (2015)
10. Suresh, A.: A study on fundamental and technical analysis. Int. J. Market. Finan. Serv. Manage. Res. **2**(5), 44–59 (2013)
11. Decker, C., Russell, R., Osuntokun, O.: eltoo: a simple layer2 protocol for bitcoin . White paper: https://blockstream.com/eltoo.pdf
12. Huang, J.Z., Huang, W., Ni, J.: Predicting Bitcoin returns using high-dimensional technical indicators. J. Finan. Data Sci. **5**(3), 140–155 (2019)

# Technology and Business Innovation

# Some Insight into Designing a Visual Graph-Shaped Frontend for Keras and AutoKeras, to Foster Deep Learning Mass Adoption

Vasile Georgescu(✉) and Ioana-Andreea Gîfu

Department of Statistics and Informatics, University of Craiova, Craiova, Romania
vasile.georgescu@edu.ucv.ro

**Abstract.** The aim of this paper is to provide some insight into designing a visual graph-shaped frontend for Keras and AutoKeras, two flagship deep learning software platforms. We also placed this particular endeavor in the larger context of deep learning mass adoption, which is just happening in many application fields, pointing out what challenges it is facing. There are several underlying conditions for going on quickly: automating the end-to-end machine learning pipeline, continuing the advances in GPU technology for supporting computing speed-up and parallelization, using tensor-enabled and GPU compatible mathematical libraries, and designing versatile GUIs capable of visually representing the network configuration in shape of a directed acyclic graph. All these blocks must be hierarchically stacked. Designing a Graph-Shaped frontend for deep learning platforms is the last, but equally essential brick to this construction and finally was the motivation behind this paper. Tensor-enabled libraries such as TensorFlow and the recent Nvidia GPU technology are the foundation layer. Keras successfully attempted to simplify the access to TensorFlow, while AutoKeras adds the automation support on top of Keras. Our proposed frontend, called Visual Keras&Autokeras, is attempting to visually emulate all the APIs related to Keras Functional model and AutoKeras AutoModel, in a codeless environment.

**Keywords:** Deep learning pipeline automation · Deep learning mass adoption · Deep learning codeless environment · Visual graph-shaped frontend design

## 1 Introduction

Deep Learning (DL) is a more recent and currently very noticed sub-branch of Artificial Intelligence (AI) and it relies on neural networks with multiple layers of hidden neurons, designed to extract higher level features from input data. DL models are capable of learning from large scaled data, forming non-linear relationships and recurrent structures. Since Finance is hugely complex, representing one of the most computationally intensive fields, with a plethora of nonlinear factors influencing each other, it appears to be a promising field of applications for DL algorithms.


M. d. P. Rodríguez García et al. (Eds.): XX SIGEF 2021, LNNS 384, pp. 133–152, 2022.
https://doi.org/10.1007/978-3-030-94485-8_11

Designing a graph frontend on top of AutoKeras [1, 2] was an appealing choice because its explicit purpose is to automate DL tasks both in terms of model choice and hyperparameter optimization. However, our experience accumulated with the "externalization" of AutoKeras' APIs in the form of a visual (codeless for users) graph frontend application raised the question of how dark a gray box should be. Indeed, AutoKeras was built on top of Keras [3] with the aim of abstracting the user from Keras details and thus gaining a higher level of manageability and ease of use. It pays for this the price to hide behind-the-scenes operations, which is normal when it comes to making optimal automatic choices. This is great for novice deep learning practitioners, but may create dissatisfaction for the most experienced ones. It is always difficult to reach the right trade-off between a high level of automation and the freedom to control significant details. So, we finally decided to further integrate Keras in our project, along with AutoKeras. This comes with additional challenges because the visual "externalization" of Keras' APIs involves explosive detailing in frontend design as well.

We found that these kinds of questions are also underlying aspects of another more fundamental question: Is Deep Learning (the most recent and advanced branch of Machine Learning) poised for mass adoption? Deep learning has been recently extended in such numerous and varied application fields that it urges us to be more reflective of the challenges of mass adoption of this new technology and its impact on the end-user. How does interference from so many radically different areas of expertise affect the proper use of these technologies in computational finance, for example? How educated should this use be and what skills does it require from an end-user without expertise in Computer Science? The concerns are both conceptual, formal, but also purely technical, because the theoretical and formal complexity of these techniques is doubled by the extreme technicality of their implementation, given the huge amount of structurally complex APIs to be explored in order to relate to the back-end application, when developing a specific new front-end application. Keras addressed this issue in an attempt to simplify the access to TensorFlow APIs [4], and AutoKeras went a step further by providing automation on top of Keras. Our aim is therefore to put the last brick to this construction, by leaving the opportunity to use DL methods in a codeless environment even to non-programmers.

Nowadays, AI in business is rapidly becoming a commonly-used competitive tool. The emergence of deep learning is just a new stage of this continuously evolving scientific and technological ecosystem. Tech giants like Google, Amazon, Facebook and Microsoft have placed huge bets on Artificial Intelligence and Machine Learning and are already using it in their products. They have set up massive cloud platforms to enable other businesses to run their machine learning models at scale. The adoption of these plethora of new technologies is spreading quickly and produces fundamental structural changes both at the organizational and societal level. Mainstream companies can also tap into cloud-based AI offerings to quickly adopt and leverage cognitive technologies without the need to invest in in-house capabilities.

However, how wide the footprint of this ecology of high-tech solutions and innovations could be? Radical innovation often raises barriers to its adoption and is naturally stressful because it introduces new standards of competitiveness that are difficult to achieve. Crossing these barriers is hampered by the obvious gap between highly skilled human resources available in academia and big high-tech pioneers on the one hand, in contrast with other organizations, including economic and financial ones, mid-tier enterprises, NGOs, governmental bodies and start-ups, on the other hand. The ways to overcome this gap apparently also come from the high-tech companies, the leading pillars of the innovative ecosystem, and consist of new solutions and services to increase the accessibility of new technologies. However, the end-user's propensity to adopt such platforms critically depends on the costs of purchasing a proprietary software, in contrast to the alternative of using an open-source software, if available.

The emergences of Automated Machine Learning Platform and Versatile GUIs are most suitable to help abstracting up from the underlying complexity of the new computational techniques, and thus to create a wider footprint of deep learning capable organizations. Moreover, their addressability can be also extended to a large variety of users, even to those who aren't deep learning experts, to nonetheless use deep learning methods to solve problems. A new growing category of machine learning and deep learning practitioners is going to emerge this way, in contrast to the researchers and academics that have dominated the field until now.

For processing data, the technology boost came from an unexpected source. High performance GPUs that were designed for gaming had the high parallel processing capability needed by machine learning algorithms. These GPUs are now widely accessible and frameworks like Google's TensorFlow [4] come with built in support for the GPU hardware.

The nowadays availability of massive data stores, and open data from many sources, in combination with the development of automated Deep Learning platforms, new versatile user interfaces, increasing power of GPUs, GPU clusters and cloud computing environments are the important pillars for a genuine mass adoption of these new technologies. Without an endeavor to fulfil these prerequisites, it would be impractical for companies to survive in the near future. Actually, the competitive race is one important source to ensure mass adoption.

The mass adoption of deep learning techniques in computational finance is an open way and a huge opportunity to advance towards more effective methods to model and forecast very nonlinear and dynamically erratic processes, with deep structural complexity. The top candidates are, of course, the STOCK and FOREX markets, currently impacted by automated and algorithmic trading technologies, thus becoming one of the largest sources of high-frequency data. The topics of interest in computational finance covers large fields of applications that started with financial time series forecasting and are quickly expanding to algorithmic trading, credit risk assessment, portfolio management, fraud detection, asset pricing, derivatives market and so on. The wider adoption of algorithmic and high-frequency trading is a natural evolutionary path in market development, driven by technological advances.

In this paper we provide some insight into designing a platform that enables end-user codeless implementations of generic deep learning applications and supports automated model choice as well as hyperparameter optimization.

The remainder of this paper is organized as follows. In the next section we take a brief look into Keras and AutoKeras functionality and the main aspects of the automation of machine learning, with special emphasis on network morphism-based Neural Architecture Search used by AutoKeras. Finally, we will present some design aspects and user-centric capabilities of our visual Graph-Shaped frontend for Keras and AutoKeras, which we chose to call Visual Keras & AutoKeras.

## 2 Deep Learning Platforms and Technologies

Deep learning platforms tends to integrate into each other hierarchically, according to their level of abstraction.

***TensorFlow*** [4] is, by far, the most widely adopted *low-level* API that serves as a *back end* for other higher-level deep learning platforms. It was developed and open sourced by members of the Google Brain team and supports deployment across CPUs, NVIDIA CUDA–compatible GPUs, and mobile and edge devices as well. Since November 2015 when it was released, TensorFlow saw an impressive increase in its adoption within the industry. TensorFlow provides highly optimized libraries for mathematical operations with tensors, which are formulated based on graphs. Typical operations such as matrix multiplications are carried out very efficiently in parallel on GPUs, during the learning process. The high density of the cores makes the GPU highly parallel, compared to traditional CPUs. GPUs are also optimized for floating-point vector operations. For this reason, GPUs are ideal for large neural networks in which typical operations such as multiplication and addition of vectors are performed for several neurons at once [17, 18].

Many DL platforms use TensorFlow as back end. However, ***Keras*** [3] is the most popular high-level DL framework that provides a second-level abstraction to DL model development, based on TensorFlow. This means that the code written in Keras acts as a wrapper that abstracts away the details of TensorFlow's APIs, which then runs on a compute instance. Keras was developed by François Chollet, now a Google engineer. The high-level Keras' API extends TensorFlow with a new level of abstraction. It contains a large number of building blocks that can be easily used and reused. Theoretically, Theano (developed by the University of Montréal) could also be used as a backend for Keras, but this solution has not been further developed since the end of 2017, because of Google's competitive advantage. Since Keras is written in Python, it has a larger community of users and supporters that feel comfortable with the combination of simplicity and flexibility it provides, while still being a high-level API.

There are two types of model APIs in KERAS: *Sequential model* API, and *Functional model* API. The *Sequential model* API is well suited to handle models with linear topology. An instance of the Sequential class is first created and a sequence of model layers are then added to it. However, a model with such a structure has some limitations. It is not easy to define models that may have multiple different input sources, produce multiple outputs or models that re-use layers. The *Functional model* API is a more flexible alternative that allows handling models with non-linear topology, shared layers, and even multiple inputs or outputs. This is motivated by the fact that a deep learning model is usually a *directed acyclic graph* (*DAG*) of layers. So, the functional API is a way to build *graphs of layers*.

The next step in designing a deep learning platform is to provide ***automation***, which is the key into utilizing the data science resources as best as possible. The multitude and high complexity of the steps involved in the Machine Learning process induce challenges that can be overcome by practitioners with a solid background in machine learning. For non-experts, instead, they may accumulate to a significant hurdle. A primary challenge is to make the data amenable for machine learning, which requires the application of a series of appropriate techniques such as data pre-processing and feature engineering. A subsequent challenge, but equally critical, is to perform algorithm selection and hyper-parameter optimization for finding the model with the best predictive performances among a multitude of candidate models. Data pre-processing usually consists of data cleaning, data normalization, column type detection (Boolean, discrete, continuous, text, etc.), column intent detection (predictors vs. labels) and task detection (classification vs. regression). Feature engineering refers to task such as: feature selection, feature extraction, meta learning, transfer learning, detection of skewed data and/or missing values). Model selection is a data-driven process that normally represents a trade-off between accuracy or goodness of fit, on the one hand, and simplicity (as stated by Occam's razor), on the other hand. It closely relates to other automatization steps, such as: hyperparameter optimization, pipeline selection under various constraints, evaluation metric selection, validation criterion selection, misconfiguration detection. Hyperparameter optimization is a critical part of deep learning since configuring neural networks is not a straight-forward task and the number of parameters that need to be tuned may be significant. Among these parameters we can mention: batch size and training epochs, learning rate and momentum, activation functions, number of neurons in the hidden layers, network weight initialization, training optimization algorithm parameters, dropout regularization, etc.

***Automated Machine Learning*** (***AutoML***) refers to techniques for automatically discovering the best-performing model for a given dataset. Automating the end-to-end process of applying machine learning covers the complete pipeline from the raw dataset to the deployable machine learning model. The goal of AutoML as an end-to-end application of machine learning is to produce simpler solutions, faster creation of those solutions, and ultimately to find a top-performing model for a specific data-driven task, such as classification or regression, that often outperform the corresponding hand-designed models. Finally, AutoML may also face the problem of supporting the end-user with the analysis and interpretation of results. The high degree of automation in AutoML allows people with limited machine learning background knowledge to use machine learning models and techniques easily. However, the automation process is not complete without providing ***graphical user interfaces*** (GUIs) and ***visualization tools***.

In response to these challenges, Google developed ***Cloud AutoML*** [19], a proprietary software described as "a suite of machine learning products that enables developers with limited machine learning expertise to train high-quality models specific to their business needs, by leveraging Google's state-of-the-art transfer learning, and Neural Architecture Search technology" [20].

On January 17, 2018 Google announced ***AutoML Vision***, "a new service that helps developers, including those with no machine learning expertise to build custom image recognition model, without having to code" [21].

***AutoKeras*** [1, 2] is an *open-source* alternative to Google's AutoML, as well as a *locally deployable system*, which thus avoids using cloud services that are not free. Moreover, it avoids facing problems such as security and privacy of the data. *AutoKeras* is developed by DATA Lab for automated machine learning at Texas A&M University and community contributors and is focusing on the deep learning tasks, instead of traditional shallow neural networks, as opposite to other automated machine learning platforms, such as Auto-Sklearn, Auto-WEKA, TPOT, etc.

AutoKeras is an implementation of the *Automated Machine Learning* concept *on top of Keras* for *automating deep learning processes.* It is an *open-source library* that provides yet friendlier APIs than Keras, while allowing a backstage access to its internal capabilities of Keras, via the tf.keras API provided by TensorFlow 2. Additionally, it allows searching for an efficient neural architecture and optimized hyperparameters. Actually, AutoKeras enables automating typical tasks such as image classification/regression, structured data classification/regression and text classification/regression and returns the model that achieves the best performance on a given dataset, selected from a number (specified by user) of models to try, under the configured constraints. This results in a TensorFlow 2 Keras (tf.keras) model, instead of a Standalone Keras model. Autokeras also provides a way of customizing the model architecture via the *Automodel API*, which allows *directed acyclic graph* of *blocks*, in a similar fashion as the *Keras' Functional IPI* does this with *Keras layers.*

AutoKeras runs in parallel on CPU and GPU, with an adaptive search strategy for different GPU memory limits (Fig. 1). See [1] for more details.

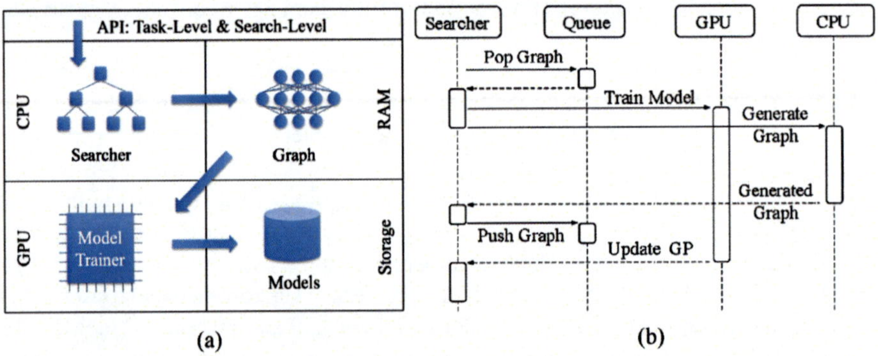


**Fig. 1.** (a) AutoKeras System Overview; (b) CPU and GPU Parallelism.

***Neural architecture search*** (NAS) [23] is a widely spread algorithm for automatically tuning deep neural networks in order to search for the best neural network architecture for the given learning task and dataset. Methods for NAS can be categorized according to the search space, search strategy and performance estimation strategy and can be underpinned with different methods, such as reinforcement learning, evolutionary algorithms, multi-objective metaheuristics, super-network-based search and so on. However, NAS techniques are known to be computationally expensive, which is a major inconvenience.

AutoKeras was designed to make use of network morphism in NAS, with the benefit of keeping the functionality of a neural network while changing its neural architecture. Although Bayesian optimization has often been used previously to search among different combinations of hyperparameters (hyperparameter tuning), it had however to be readapted when used with AutoKeras to face the more challenging case of network morphism-based NAS. As opposed to the common case of a Euclidean search space that is normally used with Bayesian optimization, the goal now is to find a node in a tree-structured search space, where each node represents a neural architecture and each edge is a morph operation [1].

Designing a versatile GUI on top of an automated Machine Learning tool can be finally seen as a way to deliver Machine Learning to non-programmers, without the need for them to write any line of code. A versatile GUI solution would be a drag-and-drop approach without coding, based on a visual (dataflow or diagrammatic) programming language, capable to manipulate program elements graphically. Since the neural network architecture can be generally represented as a directed acyclic graph, the natural choice is to use a node editor as a tool to design the GUI.

In the next section we will provide some insight into designing a Visual Graph-Shaped Frontend for Keras and AutoKeras.

## 3 Visual Keras&AutoKeras: The Frontend of an End-to-End Deep Learning Platform

### 3.1 Versatile GUIs and Visual Programming

Versatility and ease of use are the two crucial features of any GUI technology. Nowadays there is an increasing variety of open-source libraries for creative, visual, generative and data-driven projects. Two such technologies are best suited to build a versatile GUI: Node editors and Parameter trees.

Parameter trees are a system for handling hierarchies of parameters while automatically generating a GUI to display and interact with the parameters that are both editable and expandable. As an illustration, let us consider Fig. 2 that represents a relic of an older attempt to design a simple GUI for AutoKeras. It is flexible enough when the main purpose is to edit parameter values in a selective and expandable way. To make this possible, a combo-box-like button can be inserted in the tree, which allows extending the tree selectively. For instance, the "Add" button in the initial state of the tree shown in the left side of Fig. 2 can be used to choose a TextClassifier model, and thus, to expand the tree with it, as is shown in the right side of the figure.

However, although Parameter trees may allow rapid prototyping, they are limited to representing information flows that are linear or hierarchical, but fail in representing graph-like structured information. It fails, for instance, in the attempt to configure a model with nonlinear structure via the AutoModel API. The right solution to this issue is using a Node Editor.

Actually, Node Editors are the standard tools for the industry of Visual Programming, which uses a visual interface as an abstraction that is one or more levels above textual code. The tool behind it, namely the Node Editor, gives a representation of the program in the form of rectangular shapes (called nodes), connected by wires. The notable thing is that these connected nodes are topologically equivalent to a graph and allow representing data flows and data-processing pipelines. Neural networks are nothing but computational

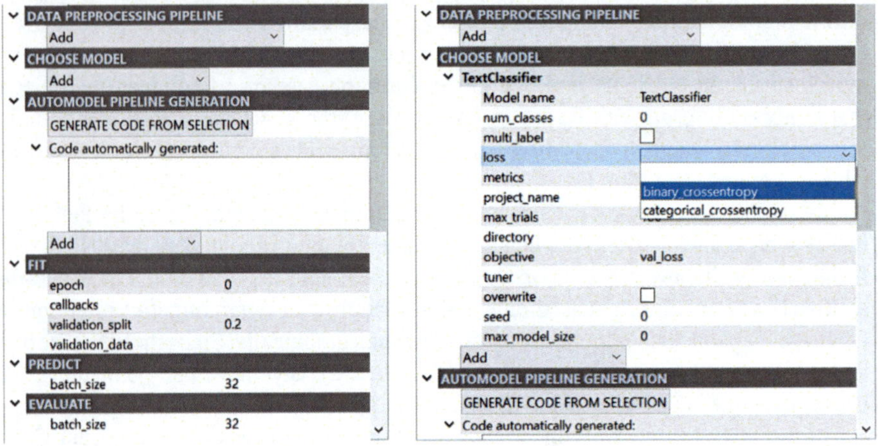


**Fig. 2.** Expandable Parameter Trees.

graphs and thus they are most likely to be visualized with nodes and wires. Based on it, programs can be created by connecting "blocks" of code. Each node is a self-contained piece of functionality allowing, for instance, to load or download datasets, pre-process the data, configure, train and validate models, and make predictions. Nodes have sockets and edges allowing them to receive and send data via connections and to update all children nodes if their value changes.

For our specific purpose, we opted to use Node Editor, a recently released open-source software, created by Pavel Křupala [23]. It is based upon the Node Editor written in PyQt5. It provides full framework for creating customizable graph, nodes, sockets and edges, full support for undo/redo and serialization into Json files, support for implementing evaluation logic, hovering effects, dragging edges, cutting lines, etc.

### 3.2 A Quick View on Keras and AutoKeras APIs

The *Functional API in Keras* [24] and the *AutoModel API in AutoKeras* are similar in nature. They both allows to deal with models that have nonlinear topology taking the form of a directed acyclic graph. Actually, they act as a function composition operator: $(f_n{}^\circ \ldots {}^\circ f_2{}^\circ f_1)(x) = f_n(\ldots f_2(f_1(x)))$. Although we can accumulate the information along linear or nonlinear sequences of nodes in the nested form given above, we still need to assign unique names to discriminate between those layers in Keras, or blocks in AutoKeras, that are used repeatedly along the same pipeline. A suffix is thus added to the layer or block name, and the function composition sequence becomes: $y_1 = f_1(x), y_2 = f_2(y_1), \ldots, y_n = f_n(y_{n-1})$.

*Models* are defined by *creating instances of layers/blocks* and *connecting* them directly to each other. Some layers act as *inputs* and *outputs* of the model. Both the Functional and AutoModel APIs enable defining a standalone *Input layer/block* that specifies the shape of input data. Keras has built-in layers for ***processing input objects*** such as image, text, structured data or other kinds of sequences like time series, audio records, etc.

After creating all model layers/blocks and connecting them together, the model can be *specified*, *trained* (*fit*), *evaluated* and used to *make predictions*.

The ***loss function*** is the metric that essentially measures the loss from the target. Popular loss functions for **regression** are: Mean Squared Error (MSE), Mean Absolute Error (MAE), Mean absolute percentage error (MAPE), Mean square logarithmic error (MSLE). For classification, used in case of categorical outcomes, we need to quantify the outcome of the class as probability and define losses based on the probability estimates as predictions. Popular choices in Keras for loss functions are binary cross-entropy and categorical cross-entropy.

***Metrics*** are functions to assess the performance of the model on a validation dataset, that is not used in training and with respect to the training dataset, but to validate the test results. Examples of functions defining metrics in Keras are: keras.metrics.binary_accuracy, keras.metrics.caetogrical_accuracy, -keras.metrics.sparse_categorical_accuracy, etc.

The ***optimizer*** is the most important part of the model training. Among the optimization algorithms available in Keras, the most used are Stochastic Gradient Descent (SGD) and Adaptive Moment Estimation (Adam). Some alternatives to them are: Adagrad, Adadelta, RMSProp, Adamax, Nadam.

The ***model configuration*** process in Keras is carried out with the **'compile'** command, for which we should provide an optimization function, a loss function, and a metric.

Once the model has been configurated, we can go ahead to the next steps: training the model using a training dataset and eventually providing a validation dataset to evaluate whether the model is well performing after each epoch. A ***fit function*** is provided by Keras to perform this step.

La last two steps are to ***evaluate*** the model and ***make predictions***.

Since Keras is the backend for Autokeras, the latter inherits the features of the former, but simplifies the user interface and adds automatic hyperparameter optimization and model selection capabilities.

AutoKeras has seven ***APIs for models***: ImageClassifier, ImageRegressor, TextClassifier, TextClassifier, TextRegressor, StructuredDataClassifier, StructuredDataRegressor, AutoModel.

The usage of ***AutoModel API*** is similar to the functional API of Keras. Basically, using it is equivalent to building a graph, whose edges are blocks and whose nodes are intermediate outputs of blocks. The pipeline starts with one of the four ***input nodes***: *ImageInput, Input, StructuredDataInput, TextInput*. There are also two ***output nodes*** that indicate what kind of task is intended to be done: *ClassificationHead* and *RegressionHead*. The search space can then be customized by adding various fine-grained blocks between the input node and the output node. Such **customizing blocks** could be: *ImageAugmentation, Normalization, TextToIntSequence, TextToNgramVector, CategoricalToNumerical, ConvBlock, DenseBlock, Embedding, Merge, ResNetBlock, RNNBlock, SpatialReduction, TemporalReduction, XceptionBlock, ImageBlock, StructuredDataBlock, TextBlock*.

### 3.3 Some Insight into Visual Keras&AutoKeras Design and Functionality

Visual Keras&AutoKeras offers a full implementation of AutoKeras capabilities, but limits itself to the implementation of only the *Keras' Functional API*, because its capabilities overlap and largely extend those of the Sequential API and hence makes it functionally redundant.

First, we will demonstrate the functionality of our visual frontend for Keras, more specifically, its capability to manipulate complex graph topologies. In order to do this, we chose to consider state-of-the art examples provided on the Keras' official site, at the section Developer guides/The Functional API. We skip the simple examples for more complicated ones.

The first case shows how to use the functional API in order to manipulate multiple inputs and outputs. This cannot be handled with the Sequential API. The purpose is to build a system for ranking customer issue tickets by priority and routing them to the correct department. The model will have *three inputs*:

- the title of the ticket (text input),
- the text body of the ticket (text input), and
- any tags added by the user (categorical input),

  as well as *two outputs*:
- the priority score between 0 and 1 (scalar sigmoid output), and
- the department that should handle the ticket (softmax output over the set of departments).

Figure 3 shows the graph edited in Visual Keras&Autokeras that allows configuring the model. Additionally, we superposed the image of the model's graph generated with the node *plot_model*, inserted in the graph, when running the configuration.

The output of each nod can be displayed via a *Display* node. For instance, the output of the node *k.Model* is a dictionary with all the information necessary to convert the visual graph representation into Python code by which to call the corresponding Keras API's. The dictionary returned by this particular node contains a sub-dictionary of all used layers, a list of inputs, a list of outputs and the associate Keras model:

```
{'layers':
  {'Input1': 'keras.layers.Input(shape=(None,), name="title")',
  'Embedding1': 'keras.layers.Embedding(10000,64)(Input1)',
  'LSTM1': 'keras.layers.LSTM(128)(Embedding1)',
  'Input2': 'keras.layers.Input(shape=(None,), name="body")',
  'Embedding2': 'keras.layers.Embedding(10000,64)(Input2)',
  'LSTM2': 'keras.layers.LSTM(32)(Embedding2)',
  'Input3': 'keras.layers.Input(shape=(12,), name="tags")',
  'Concatenate1': 'keras.layers.Concatenate()([LSTM1,LSTM2,Input2])',
  'Dense1': 'keras.layers.Dense(1, name="priority")(Concatenate1)',
  'Dense2': 'keras.layers.Dense(4, name="department")(Concatenate1)'},
 'inputs': '[Input1,Input2,Input3]',
 'outputs': '[Dense1,Dense2]',
 'model': 'keras.Model(inputs,outputs)'}
```

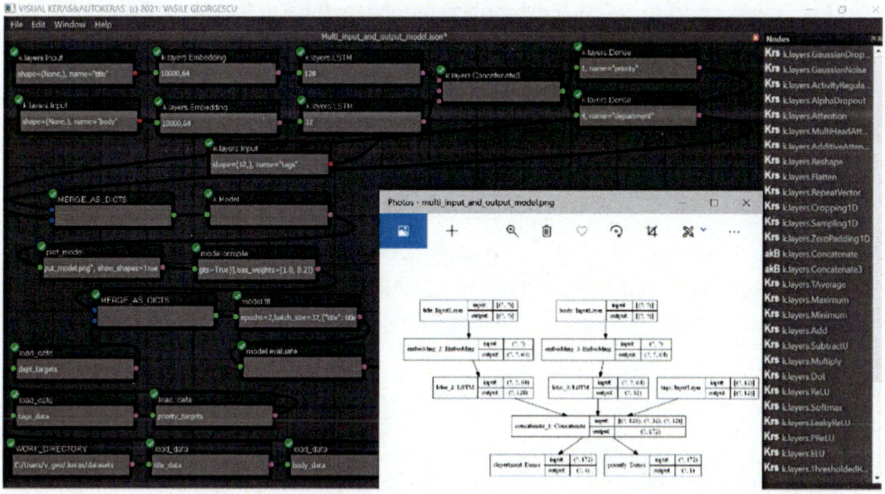


**Fig. 3.** The corresponding visual graph representation of a multi-input and multi-output Keras model that is then passed to the *Keras Functional API* to be configured. The image of the model's graph has been stored as a *png* file via the node *plot_model*.

Note that the lists of inputs and outputs need not to be specified in the node k.model, but are instead collected automatically from the information flow. The code in Python that is automatically generated from this dictionary is like this:

```
Input1 = keras.layers.Input(shape=(None,), name="title")
Embedding1 = keras.layers.Embedding(10000,64)(Input1)
LSTM1 = keras.layers.LSTM(128)(Embedding1)
Input2 = keras.layers.Input(shape=(None,), name="body")
Embedding2 = keras.layers.Embedding(10000,64)(Input2)
LSTM2 = keras.layers.LSTM(32)(Embedding2)
Input3 = 'keras.layers.Input(shape=(12,), name="tags")
Concatenate1 = keras.layers.Concatenate()([LSTM1,LSTM2,Input2])
Dense1 = keras.layers.Dense(1, name="priority")(Concatenate1)
Dense2 = keras.layers.Dense(4, name="department")(Concatenate1)
inputs = [Input1, Input2, Input3]
outputs = [Dense1, Dense2]
model = keras.Model(inputs, outputs)
```

Figure 4 also shows a second pipeline that is used to feed data to the model. We generated the data in the same way as in the provided example, and then stored it as files in Numpy *npz* format on the local disk. The first node when loading data from the local disk is always "WORK_DIRECTORY". It also precedes any node for downloading data via URLs, or via financial API (e.g., Yahoo, Kaggle, etc.), or other kind of public API available. The information editable in such nodes refers only to the file name or subdirectory name, not the full path.

Let as now consider another example provided on the Keras' site, at the same section, whose aim is to manipulate non-linear connectivity topologies via the functional API, i.e., to configure models with layers that are not connected sequentially. Actually, a toy ResNet model for CIFAR10 is built to demonstrate the use of residual connections. The code in Python is presented below:

```
inputs = keras.Input(shape=(32, 32, 3), name="img")
x = layers.Conv2D(32, 3, activation="relu")(inputs)
x = layers.Conv2D(64, 3, activation="relu")(x)
block_1_output = layers.MaxPooling2D(3)(x)
x = layers.Conv2D(64, 3, activation="relu", padding="same")(block_1_output)
x = layers.Conv2D(64, 3, activation="relu", padding="same")(x)
block_2_output = layers.add([x, block_1_output])
x = layers.Conv2D(64, 3, activation="relu", padding="same")(block_2_output)
x = layers.Conv2D(64, 3, activation="relu", padding="same")(x)
block_3_output = layers.add([x, block_2_output])
x = layers.Conv2D(64, 3, activation="relu")(block_3_output)
x = layers.GlobalAveragePooling2D()(x)
x = layers.Dense(256, activation="relu")(x)
x = layers.Dropout(0.5)(x)
outputs = layers.Dense(10)(x)
model = keras.Model(inputs, outputs, name="toy_resnet")
```

Figure 4 shows how we can emulate this code by means of the Visual Keras&AutoKeras graph editor.

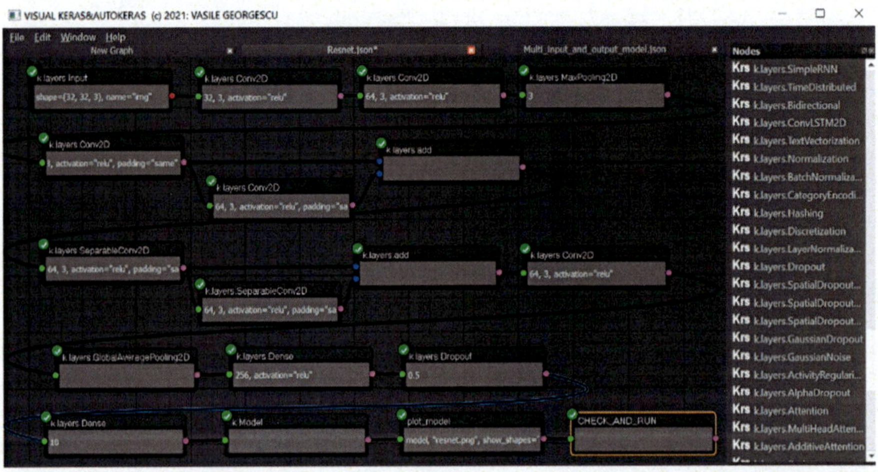


**Fig. 4.** Emulating the code of the toy RESNET example, via the Visual Keras&AutoKers graph editor.

The graph representation of the model is absolutely similar with that obtained with the Keras code in Python. Our node "plot_model", which is the graphical interface for keras.utils.plot_model, generates the graph in the Fig. 5, which has been split because it is too long.

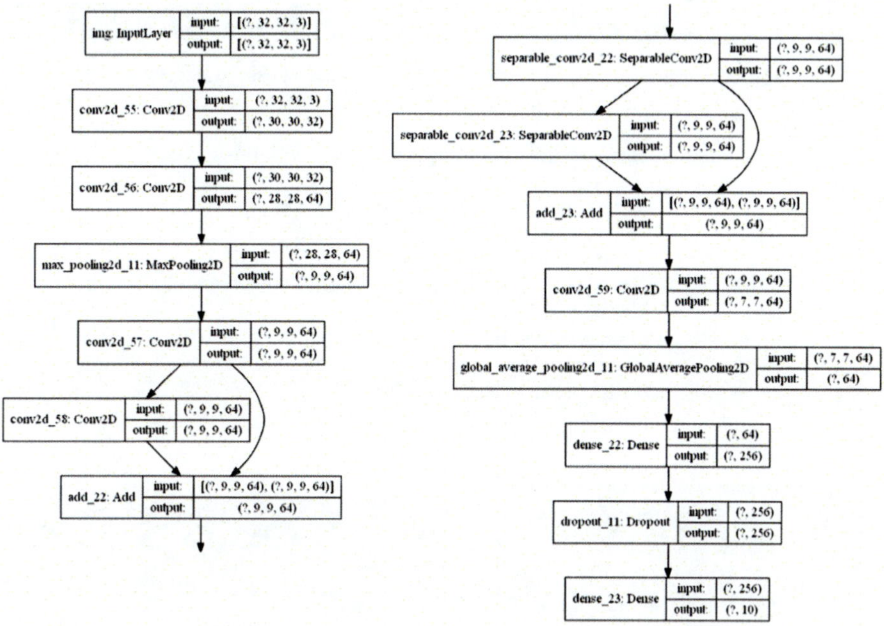


**Fig. 5.** Graph representation of the toy ResNet model.

Next, we will demonstrate the functionality of our visual frontend for AutoKeras, and we also chose to consider state-of-the art examples provided on the AutoKeras' official site.

As we have already mentioned, AutoKeras has seven ***APIs for models***: ImageClassifier, ImageRegressor, TextClassifier, TextClassifier, TextRegressor, StructuredDataClassifier, StructuredDataRegressor, AutoModel. The *AutoModel API* is functionally similar to the *Keras' Functional API*. All the capabilities of AutoKeras can be emulated with Visual Keras&AutoKers. We will demonstrate some of them, by replicating in the visual environment the following sequences of code (Figs. 6, 7, 8, 9, 10, 11, 12, 13, 14):

```
# 1. Image Classification using for MNIST dataset using ImageClassifier
clf = ak.ImageClassifier(
        overwrite=True, max_trials=1)
clf.fit(x_train, y_train, validation_split=0.15, epochs=10)
```

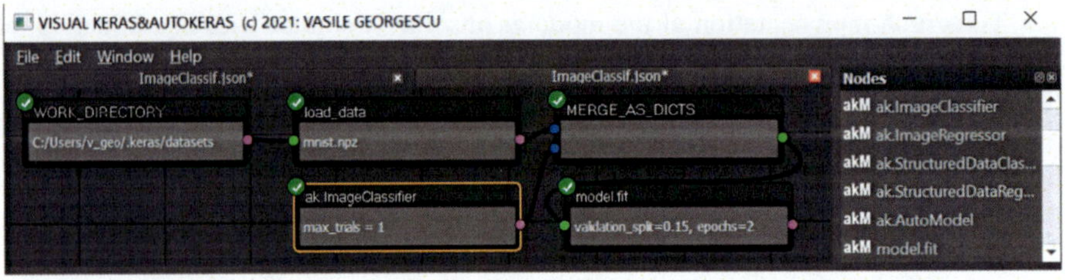


**Fig. 6.** Image Classification using for MNIST dataset using ImageClassifier.

```
# 2. Image Classification using AutoModel with ImageBlock
        input_node = ak.ImageInput()
        output_node = ak.ImageBlock(
            block_type="resnet", normalize=True,
            augment=False )(input_node)
        output_node = ak.ClassificationHead()(output_node)
        clf = ak.AutoModel(
            inputs=input_node,  outputs=output_node,
            overwrite=True, max_trials=1)
        clf.fit(x_train, y_train, epochs=10)
          predicted_y = clf.predict(x_test)
```

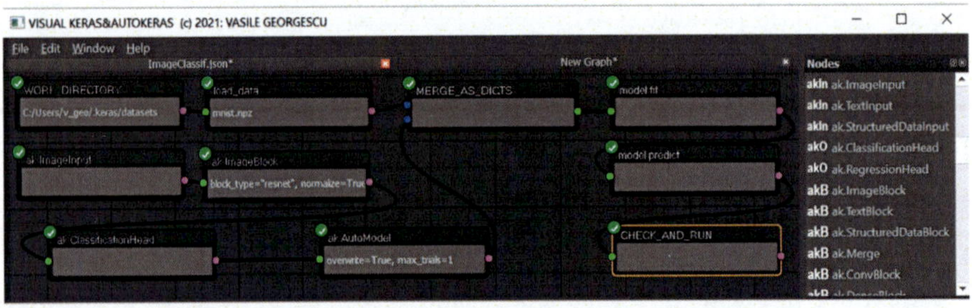


**Fig. 7.** Image Classification using AutoModel with ImageBlock.

```
# 3. Image Classification using AutoModel with customization
        input_node = ak.ImageInput()
        output_node = ak.Normalization()(input_node)
        output_node = ak.ImageAugmentation(horizontal_flip=False)(output_node)
        output_node = ak.ResNetBlock(version="v2")(output_node)
        output_node = ak.ClassificationHead()(output_node)
        clf = ak.AutoModel(
            inputs=input_node, outputs=output_node,
            overwrite=True, max_trials=5)
        clf.fit(x_train, y_train, epochs=10)
```

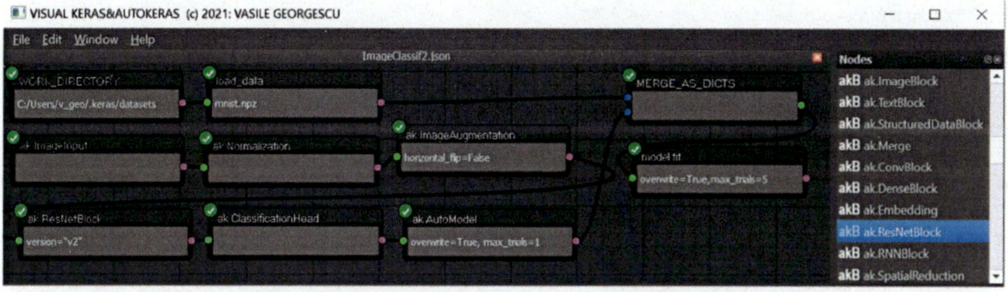


**Fig. 8.** Image Classification using AutoModel with customization.

```
# 4. Text Classification for IMDB dataset using TextClassifier
    clf = ak.TextClassifier(
        overwrite=True, max_trials=1)
    clf.fit(x_train, y_train, validation_split=0.15, epochs=2)
    predicted_y = clf.predict(x_test)
    clf.evaluate(x_test, y_test)
```

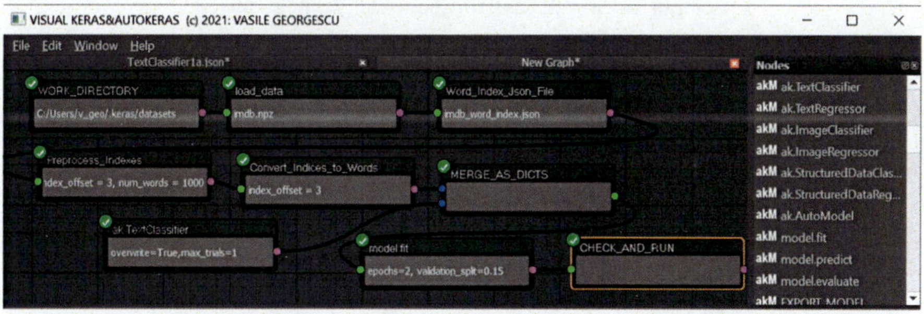


**Fig. 9.** Text Classification for IMDB dataset using TextClassifier.

```
# 5. Image Classification using AutoModel with TextBlock
    input_node = ak.TextInput()
    output_node = ak.TextBlock(block_type='ngram')(input_node)
    output_node = ak.ClassificationHead()(output_node)
    clf = ak.AutoModel(
        inputs=input_node, outputs=output_node,
        overwrite=True, max_trials=1)
    clf.fit(x_train, y_train, epochs=2)
```

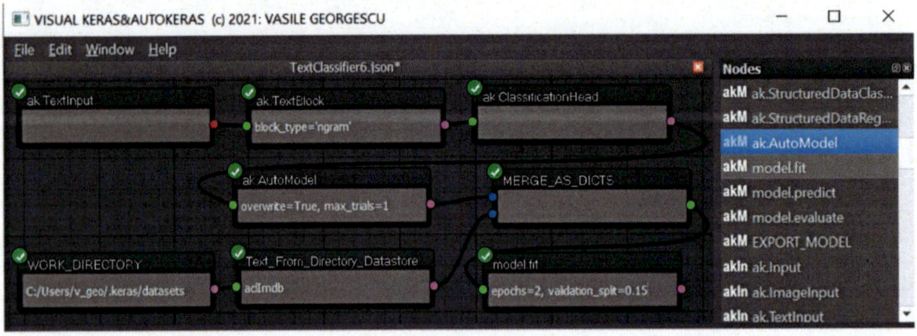


**Fig. 10.** Image Classification using AutoModel with TextBlock.

```
# 6. Text Classification using AutoModel with customization
    input_node = ak.TextInput()
    output_node = ak.TextToIntSequence()(input_node)
    output_node = ak.Embedding()(output_node)
    output_node = ak.ConvBlock(separable=True)(output_node)
    output_node = ak.ClassificationHead()(output_node)
    clf = ak.AutoModel(
        inputs=input_node, outputs=output_node,
        overwrite=True, max_trials=1)
    clf.fit(x_train, y_train, epochs=2)
```

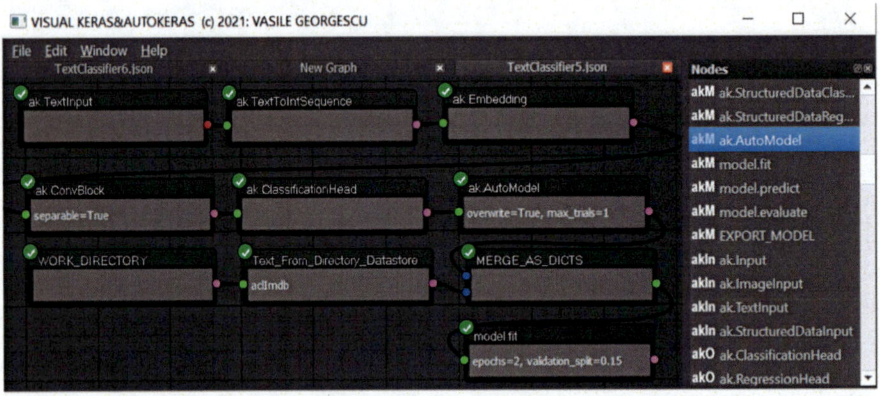


**Fig. 11.** Text classification using AutoModel with customization.

The node ak.AutoModel returns the following dictionary, which can be easily converted into Python code:

```
{'blocks':
  {'TextInput1': 'ak.TextInput()',
   'TextToIntSequence1': 'ak.TextToIntSequence()(TextInput1)',
   'Embedding1': 'ak.Embedding()(TextToIntSequence1)',
   'ConvBlock1': 'ak.ConvBlock(separable=True)(Embedding1)',
   'ClassificationHead1': 'ak.ClassificationHead()(ConvBlock1)'},
 'inputs': '[TextInput1]',
 'outputs': '[ClassificationHead1]',
 'automodel': 'ak.AutoModel(inputs,outputs)'}
```

```
# 7. StructuredData Regression for California housing dataset
        reg = ak.StructuredDataRegressor(
            overwrite=True, max_trials=3)
        reg.fit(x_train, y_train, validation_split=0.15, epochs=10)
        predicted_y = reg.predict(x_test)
        reg.evaluate(x_test, y_test)
```

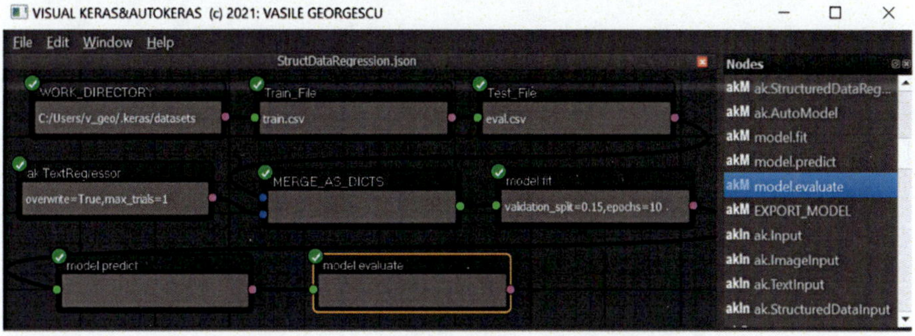


**Fig. 12.** Structured Data Regression for California housing dataset using StructuredDataRegressor.

```
# 8. StructuredData Regression using AutoModel with custumization
        input_node = ak.StructuredDataInput()
        output_node = ak.CategoricalToNumerical()(input_node)
        output_node = ak.DenseBlock()(output_node)
        output_node = ak.RegressionHead()(output_node)
        reg = ak.AutoModel(inputs=input_node, outputs=output_node, max_trials=3,
                    overwrite=True)
        reg.fit(x_train, y_train, epochs=10)
        model = reg.export_model()
        model.summary()
        model.predict(x_train)
```

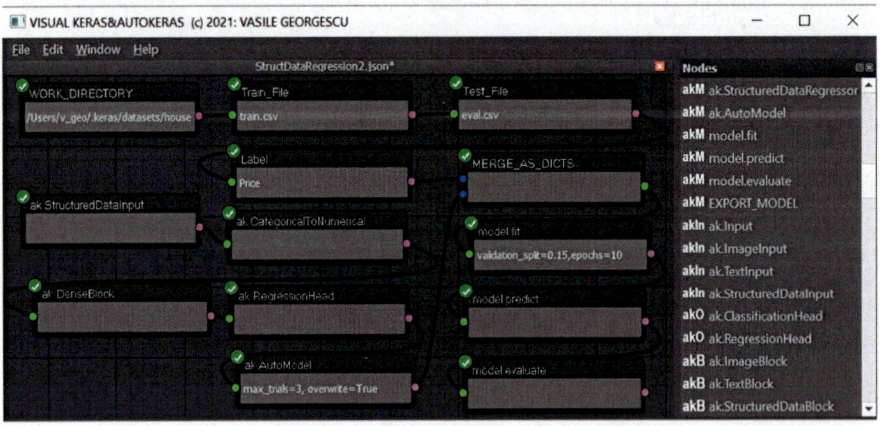


**Fig. 13.** Structured Data Regression using AutoModel with customization.

```
# 9. Configuring the model using AutoModel for a multi-input, multi-output case
    input_node1 = ak.ImageInput()
    output_node = ak.Normalization()(input_node1)
    output_node = ak.ImageAugmentation()(output_node)
    output_node1 = ak.ConvBlock()(output_node)
    output_node2 = ak.ResNetBlock(version='v2')(output_node)
    output_node1 = ak.Merge()([output_node1, output_node2])
    input_node2 = ak.StructuredDataInput()
    output_node = ak.CategoricalToNumerical()(input_node2)
    output_node2 = ak.DenseBlock()(output_node)
    output_node = ak.Merge()([output_node1, output_node2])
    output_node1 = ak.ClassificationHead()(output_node)
    output_node2 = ak.RegressionHead()(output_node)
    auto_model = ak.AutoModel(
        inputs=[input_node1, input_node2],
        outputs=[output_node1, output_node2],
        overwrite=True, max_trials=2)
    auto_model.fit(
        [image_data, structured_data],
        [classification_target, regression_target],
        batch_size=32, epochs=3)
```

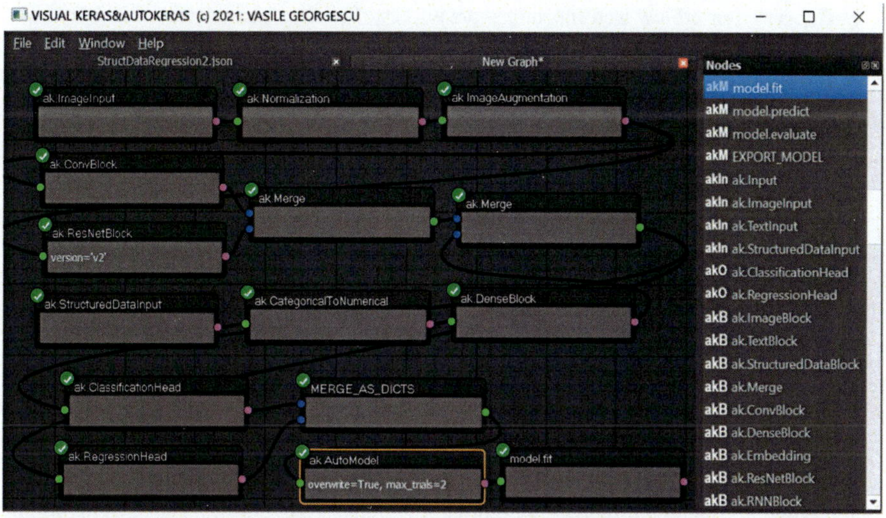


**Fig. 14.** Configuring the model using AutoModel for a multi-input, multi-output case.

# 4 Conclusions

The examples given in the previous section demonstrate the powerful capabilities of Visual Keras&Autokeras to emulate the Keras and AutoKeras APIs in a codeless environment, at any level of complexity. It uses a node editor to allow building any kind of network configurations, if they are compatible with those accepted by Keras and AutoKeras, i.e., they take the shape of a directed acyclic graph. At this current development stage, access is provided to all AutoKeras APIs, as well as to all Keras APIs relating to the Functional model. The lack of a similar implementation for the Sequential API is just a temporary choice and was motivated by the redundancy in its functionality, when comparing with the Keras Functional API. Visual Keras&Autokeras is currently under development. However, the current pre-released version behaved stable for a long-enough while now, despite some frequent additions of new features.

Although a codeless environment for using deep learning methods is a great opportunity for non-programmers, it is not a free lunch, by any means. Configuring complex networks is a deeply analytical and *task-anchored* process by its very nature. From an extended and more positive perspective, however, we can agree that automated configurations could be close to or even better than their human alternatives, provided that targets are well specified and the results receive appropriate interpretations.

# References

1. AutoKeras web site. https://autokeras.com/. Last accessed 28 May 2021
2. Haifeng, J., Qingquan, S., Xia, H.: Auto-Keras: An Efficient Neural Architecture Search System, arXiv:1806.10282 (2019)
3. Keras web site. https://keras.io/. Last accessed 28 May 2021

4. Google's TensorFlow web site. https://www.TensorFlow.org/. Last accessed 28 May 2021
5. Elman, J.L.: Finding structure in time. Cogn. Sci. **14**(2), 179–211 (1990)
6. Jordan, M.I.: Serial order: a parallel distributed processing approach. In: Neural-Network Models of Cognition – Biobehavioral Foundations. Advances in Psychology. Neural-Network Models of Cognition, vol. 121, pp. 471–495 (1997)
7. Williams, R.J., Hinton, G.E., Rumelhart, D.E.: Learning representations by back-propagating errors. Nature **323**(6088), 533–536 (1986)
8. Hochreiter, S., Schmidhuber, J.: Long short-term memory. Neural Comput. **9**(8), 1735–1780 (1997)
9. Cho, K., et al.: Learning Phrase Representations using RNN Encoder-Decoder for Statistical Machine Translation, arXiv:1406.1078 (2014)
10. LeCun, Y., et al.: Backpropagation Applied to Handwritten Zip Code Recognition. AT&T Bell Laboratories (1989)
11. Hubel, D.H., Wiesel, T.N.: Receptive fields and functional architecture of monkey striate cortex. J. Physiol. **195**(1), 215–243 (1968)
12. Krizhevsky, A., Sutskever, I., Hinton, G.E.: ImageNet classification with deep convolutional neural networks. Commun. ACM **60**(6), 84–90 (2017)
13. LeCun, Y., Bottou, L., Bengio, Y., Haffner, P.: Gradient-based learning applied to document recognition. Proc. IEEE **86**(11), 2278–2324 (1998)
14. Goodfellow, I., Bengio, Y., Courville. A.: Deep Learning, MIT Press (2016)
15. Zhang, W.: Shift-invariant pattern recognition neural network and its optical architecture. In: Proceedings of Annual Conference of the Japan Society of Applied Physics (1988)
16. Srivastava, N., Hinton, C.G., Krizhevsky, A., Sutskever, I., Salakhutdinov, R.: Dropout: a simple way to prevent neural networks from overfitting. J. Mach. Learn. Res. **15**(1), 1929–1958 (2014)
17. Hinton, G.E., Osindero, S., Teh, Y.W.: A fast learning algorithm for deep belief nets. Neural Comput. **18**(7), 1527–1554 (2006)
18. Ciresan, D., Meier, U., Gambardella, L., Schmidhuber, J.: Deep big simple neural nets for handwritten digit recognition. Neural Comput. **22**(12), 3207–3220 (2010)
19. Google's Cloud AutoML web site. https://cloud.google.com/automl. Last accessed 28 May 2021
20. Google's Cloud AutoML docs web site. https://cloud.google.com/automl/docs. Last accessed 28 May 2021
21. TechCrunch web site. https://techcrunch.com/2018/01/17/googles-automl-lets-you-train-custom-machine-learning-models-without-having-to-code/. Last accessed 28 May 2021
22. Elsken, T., Metzen, J.H., Hutter, F.: Neural architecture search: a survey. J. Mach. Learn. Res. **20**(55), 1–21 (2019)
23. Křupala P.: Node Editor web site. https://pypi.org/project/nodeeditor/#description. Last accessed 28 May 2021
24. Keras Functional API web site. https://keras.io/guides/functional_api/. Last accessed 28 May 2021
25. AutoKeras AutoModel API web site. https://autokeras.com/tutorial/customized/. Last accessed 28 May 2021

# Elasticities on a Mixed Integer Programming Model for Revenue Optimization

Jesus Lopez-Perez(✉)

Facultad de Contaduría Pública y Administración, Universidad Autónoma de Nuevo León, Av. Universidad S/N, Cd. Universitaria, 66451 San Nicolás de los Garza, NL, Mexico
jesus.lopezpz@uanl.edu.mx

**Abstract.** This paper addresses a price optimization problem to maximize the revenue for a convenience store chain based in South America while keeping control of traffic and assortment at the stores. The challenge is to design an efficient algorithm to handle large scale multi-product scenarios. The paper proposes a novel approach based on 2 major components. On the predictive component there is a collection of 5 different models using parametric and non-parametric machine learning techniques. All these vectors are processed and integrated into a final forecast using a novel Ensemble approach improving the forecasting accuracy by 5–8%. On the second component the ensemble forecasts and elasticities are processed into an optimization model. The optimization problem requires to maximize revenue as the interaction of discrete price points and multi-cross demand responses. These non-linear interactions are identified as a binary quadratic problem (BQP). BQP problems are NP-hard and computationally intractable. The paper presents an interesting approach to reformulate the original BQP problem into a Mixed Integer programming (MIP) model based on McCormick-envelope transformations. The problem was solved efficiently using an open source solver for a very large set of product-store combinations. To cope with computational time-constraints on a real-world setting the instances were solved with a time limit parameter. The experiments reported show the proposed solution is able to solve millions of product-stores combinations on short computational times. An important increase on revenue about 6% while controlling volume and assortment constraints is reported on the paper as the main contribution for the business outcome.

**Keywords:** Optimal pricing · Parametric elastic-net modeling · Non-parametric machine learning · Demand response · Ensemble modeling · Cross price elasticity · Mixed integer programming · McCormick envelopes

## 1 Introduction

Price optimization is an important research topic on the general context of marketing & economic sciences. In a broader perspective, this problem involves the use of pricing and inventory management to balance supply and demand. Particularly in retailing it is expected to synchronize pricing and promotions with an aligned assortment. Sadly, price


M. d. P. Rodríguez García et al. (Eds.): XX SIGEF 2021, LNNS 384, pp. 153–177, 2022.
https://doi.org/10.1007/978-3-030-94485-8_12

discounts and markdowns practice has been just the response to cope with uncertainty. Historically retailers have been addressing this problem statically using simple policies and rules based on the time the product is on the shelf (i.e., X% off after Y weeks and at the end tax-donation). Unplanned price discounts could be effective to address obsolete inventory, however this deviates the real focus on the business for revenue generation. Importantly, this static strategy has been occurring more because of inventory concerns rather than a rational effort of maximize the business revenue and optimize the assortment. Our claim is that this static approach should be replaced with a more educated process to consistently maximize revenue and profit using a "value-based" pricing approach to maximize the amount a customer is willing to spend. To illustrate this, it is the case of Sears Holdings Corporation which reported in 2008 a net loss of $56 million attributed to a price strategy eroding the margin [1].

Although the retailer is on control of price decisions nevertheless is the end shopper who is in control of the actual demand response. Even in some cases "strategic" customers might take control using a delayed response to trigger additional price discounts. It was until the last decade when scientific research work has been shifting the focus into the pricing process rather than the supply chain problem. The problem has been transformed into an opportunity to experiment with pricing looking to shape and optimize the demand response to maximize traffic and revenue. A recent successful business example is when Starbucks in 2020 raised their beverage prices by an average of 1% across U.S. Company's third quarter in 2020 net income rose 25% ($333 million up to $418 million). This change did not impact the price of high-margin products. Cross-price elasticity analysis was done to specific products rather than naively apply the same increase for all the products. Collaterally speaking, this is, increase the price of specific products to persuade customers to purchase other products with higher margins (i.e., customer migration).

Price elasticity is the modeling between price and demand, and it is used to calculate the demand response at different price points [2]. Price elasticity of demand is a well-known concept and defined as $E = \%\Delta D / \%\Delta P$. Changes in Prices (P) and Demand (D) are expressed in percentages from the original-baseline demand and price for the product. As expected, the lower the price the higher is the demand. However, as the price decreases the demand increases at a slower rate. After a certain price-point the increase in demand will not compensate the decrease in price and then the revenue will erode. Figure 1 illustrates the effects of changes in price on the demand response. Furthermore, demand generated from price reduction can be temporary or shifted from one product to another (i.e., cannibalization effect). These effects can confound the true impact when evaluating a price elasticity effect. As a result, determination of price elasticity is a challenging task as for each product we may have different responses. We may have highly elastic products if for a small change in price there is a large change in demand, and it is inelastic if for a change in price there is not much change in demand. As on the Starbucks example this revels an inelastic demand; therefore, a small price increase can large positive impact on revenue because demand is almost the same.

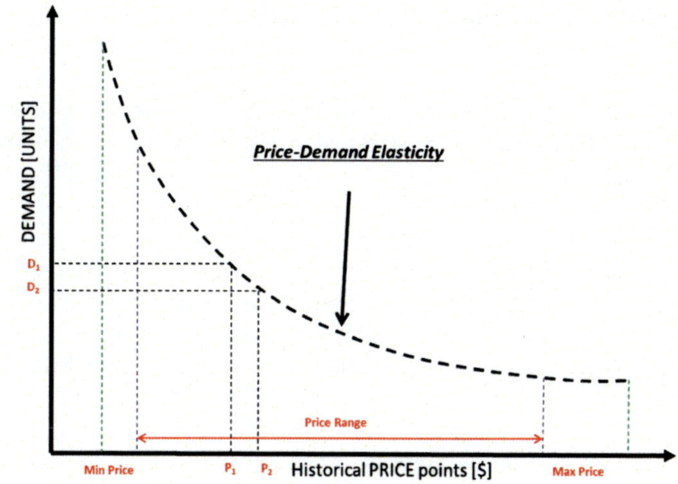


**Fig. 1.** Price Elasticity expressed as price and demand relationship.

## 2 Problem Definition and Modeling Assumptions

The problem addressed on the present paper is as a strategic pricing problem dealing with a large set of substitutable and complemental products over a finite selling horizon for a convenience store chain based in South America. An intrinsic challenge in our problem is to estimate how consumers will respond to different price-points. The price of one product can affect sales of a related product in two ways: (a) substitution is where both products effects are aligned (i.e., price increase in a product positively increase sales of other products); (b) complementary is on the other way around (i.e., price increase in a product negatively affects sales of other products). The objective of our work is to recommend optimized pricing to support the recurrent process to maximize a multi-objective function defined by revenue, profit margin and particularly the assortment configuration. Business-wise the objective is to provide a solution to optimize and facilitate the decision process by applying Advanced Data Science models. Taking in mind multiple price-points interacting with multiple elastic demand models estimated for multiple store-product combinations, then the problem is to find the best price strategy for all of these possible scenarios keeping the complexity of the solution tractable to be implemented on a real-world business setting.

Justification and business relevance. In the current problem the retailer does not have accurate demand prediction models and price decisions are given based on simple policies as discussed on Sect. 1 rather than on a formal optimization approach. Aggressive price discounts may increase traffic and volume but will erode drastically the margins. On the other hand, conservative prices may generate unsold inventory requiring deeper price discounts and loss of revenue over time. There is no have evidence to prove the existence of "strategic" customers delaying a purchase decision to take advantage of future price discounts. Either way from retailer's perspective the aim is to optimize the pricing process and discourage these kind of customer behaviors.

## 3 Literature Review and Related Work

In this section the current literature review is presented, in general for price optimization and in particular for the relevant strategic pricing use case. The price optimization process is diverse but in general can be found to design revenue structures, promotional pricing and markdown pricing in different industries including retail, airlines, hotels and financial services [3]. Also, this process can be found in more complex settings like customer targeting when the model is used to evaluate pricing effects for different customer segments. While more focused journals (i.e., Journal of Retailing, Marketing Science) regularly include papers that are useful in the revenue management context, the pricing decisions versus the assortment and inventory problems is practically not included in this discussion. On this literature there is aspects often driven by concerns like forecasting and predictive accuracy. In Gupta and Pathak [4] they have customer segments using K-means using a regression model to estimate a price range for each cluster. A logistic regression post process is presented to estimate customer's likelihood of buying the product on each cluster. In Kopalle [5] and Levy et al. [6], they include factors to consider when optimizing retail merchandising decisions. In Bertsimas and de Boer [7], they presented a deterministic approximation to the full stochastic price optimization problem excluding inventory-rationing decisions.

About "strategic-customer" problems, Shen and Su [8] and Levin et al. [9] presented a work including aspects like adoption and perceived risk. Levin et al. [9] model the problem as a stochastic dynamic game and rely on the existence of a unique subgame-perfect equilibrium pricing policy. They argue as buying decisions are purely based on rational expectation of future prices the problem should be analyzed in a game theoretic framework (i.e., Nash equilibria). In Ahn et al. [10], they present the variant to address strategic customers delaying buy-decisions. An important step is taken in Smith [11] and Valkov [12], which describe the implementation and use of price optimization systems by various companies and report some related results. However, they do not discuss the details about the methodology they used to calculate the impacts presented.

Several articles covering questions about dynamic pricing can be found in Operations Research and Management Science literature. Rather than providing an exhaustive review, this section will highlight a few studies that can provide a starting point for a more comprehensive review. Comprehensive reviews for dynamic pricing can be found in Talluri and Van Ryzin [13] also in Ozer and Phillips [14]. Bitran and Caldentey [15] provide a stochastic model providing a technical review and unified framework focused on dynamic pricing models. Maglaras and Meissner [16] provide a unifying framework for pure pricing and pure inventory problems in a multi-product stochastic dynamic setting. A common aspect of these 2 papers is that they are not explicitly including capacity rationing decisions as the models do not differentiate between actual sales and demand decisions.

Considering the discrete nature of the underlying price-points we can find some studies using mixed-integer programming [17–19]. In common these studies fail to handle large scale settings for problems with hundreds or thousands of products since computational time exponentially increases over increasing combination of products and store-locations. In Ito and Fujimaki [20] and Yabe et al. [21], they proposed a prescriptive framework to efficiently solve multi-product price optimization with non-linear demand models. However, the proposed method cannot deal with complex business constraints such as the number of discounted products. The problem is presented into a binary quadratic programming problem, and they proposed a relaxation method based on semi-definite programming (SDP). Although their approach significantly improved computational efficiency it is not sufficiently scalable for large scale problems with millions of product-stores combinations. Maddah and Bish [22] analyze static pricing for multiple competing products. A fast heuristic algorithm is designed to implement the proposed model in real-time. Caro and Gallien [17] implement a markdown multi-product pricing decision support tool for the fast-fashion retailer Zara.

Due to the underlying non-linear classes of demand functions on the literature review there is a lot of nonlinear mixed integer program (NMIP) formulations. On this regard, one of the most important modeling challenges that are addressed on the reviewed literature is about the cross-price elasticities (i.e., cross price effects between interrelated products). Interesting some linear constraints were implemented in Cohen et al. [23] to handle different type of business rules implemented to reflect the cross-price interactions between several products. In general, the difficulty is how to handle higher dimensional solution spaces as more products are considered into the cross-price elasticity predictive models. As a result, exact solution optimizers struggle to reach the global optimum [24]. Cross-price effects are not only non-linear but asymmetric as well. This is explained as how product "A" influences product "B" can be different as how product "B" influences product "A". It has been shown in Bienstock [25] that this problem is NP-hard and do not satisfy the conditions to be solved using Branch and Cut or Benders strategies [26].

The underlying idea of using recency and myopic demand models back to what we found on Besbes and Zeevi [27]. They deal with a multi-period single product pricing problem with the particularity of an unknown demand curve. Here the objective is to adjust prices in each period so as to maximize cumulative expected revenues over a given time horizon. The question they addressed is about how large the revenue loss is when relying on under-specified myopic model to predict the demand response (i.e., applying variants of simple pricing policies). This operates in a myopic manner as they apply minimal price experimentation. Surprisingly, they prove that these simple minded and incomplete models are still quite useful. Here below Table 1 includes a taxonomy about the different variants of the dynamic pricing problem discussed on our review.

**Table 1.** Classification scheme criterion.

| Concept | Taxonomy |
|---|---|
| Pricing policy | Markdowns and markups (temporary price discount) Based on competitors' price |
| Pricing decisions | Static and defined from the beginning of the selling horizon Dynamic defined during the selling horizon |

(*continued*)

**Table 1.** *(continued)*

| Concept | Taxonomy |
|---|---|
| Cross price elasticities | Own only (i.e., no assortment effects) Cross price elasticity including assortment effects |
| Demand process | Single type of products Multi-products (i.e., perishable & non-perishable) |
| Capacity | As constrained parameter (i.e., inventories & replenishments) as decision variable |
| Objective function | Single-objective based on volume or revenue Multi-objective based on revenue and profit simultaneously |

## 4 Proposed Methodology

As defined on previous section the objective of this work is to recommend an optimized pricing to maximize a multi-objective function defined by revenue, profit margin and specific assortment configuration over a finite time-horizon. Three types of variables are defined: decision, target, and predictor variables. Decision variables are those required to optimize (i.e., prices for specific product-store-week combinations). Target variables are defined to predict the demand response for incumbent, complement and substitutable products. Predictor variables are all the attributes and features utilized as causal factors to predict the demand response. The proposed methodology is outlined in 3 components.

- Component 1: Network Analysis. Identify which products have demand interactions with other products. Identify groups of highly-related products. This component was developed relying on an open-source Networkx library implementation in python https://networkx.org/ (Fig. 2).

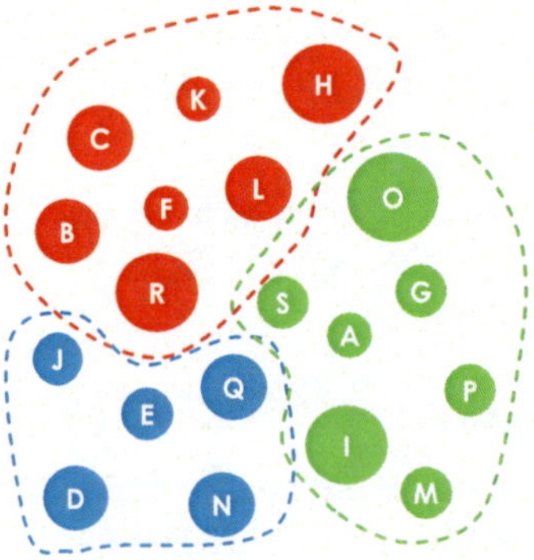


**Fig. 2.** Network analysis to model demand interactions among related products inside a category

- Component 2: Predictive modeling. Identify TRUE relationships between prices and demands across related products. Calculate accurate elasticities to estimate demand response from price changes. A collection of different predictive models using parametric and non-parametric techniques is used as the proposed approach to generate the underlying coefficients to estimate both: (a) own-price elasticities and (b) the cross-price effects when used for related products (i.e., complement and substitutable products). This component was developed relying on the open-source Sci-Kit learn library implementation in python https://scikit-learn.org/stable/.
- Component 3: Optimization modeling. Identify the optimal prices for each product to maximize the expected revenue given relevant business constraints. An advanced **optimization model** is proposed relying on the cross-price elasticities coefficients generated from the previous predictive models. This optimization problem could be defined as a concave binary quadratic problem (BQP). Therefore, the problem has a non-convex representation, and it is NP-Hard. As a result, the problem was reformulated into a linear Mixed Integer problem (MIP) using McCormick-envelope transformations. This component was developed relying on the open-source SCIP library implementation in python https://www.scipopt.org/.

The predictive modeling will be presented on Sect. 5. The exact-solution MIP formulation will be formally presented and discussed on Sect. 6. Here below in Fig. 3 it is a visual representation of the described methodology explained at a high level.

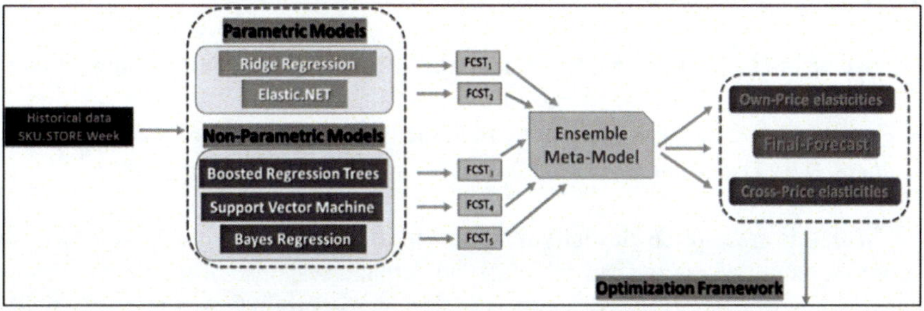


**Fig. 3.** High level proposed methodology

## 5 Predictive Modeling

As we pointed out Price elasticity is the concept to reflect how the price impacts the demand and it is used to calculate the demand response at different price points. On the other hand, cross Price elasticity is the concept used to explain how the price of a complemental or supplemental product can impact the demand of a related or incumbent product. Yes indeed, we want to identify how the price change on product B impacts the change on the demand for the incumbent product A. We can express this as follows:

$$\text{Cross-price Elasticity}: \Delta \text{ in price of Product - B} \rightarrow \Delta \text{ of demand in Product-A}$$

Figure 4 illustrates the difference on price elasticity versus cross-price elasticity of product B when impacting the demand response of the incumbent product A.

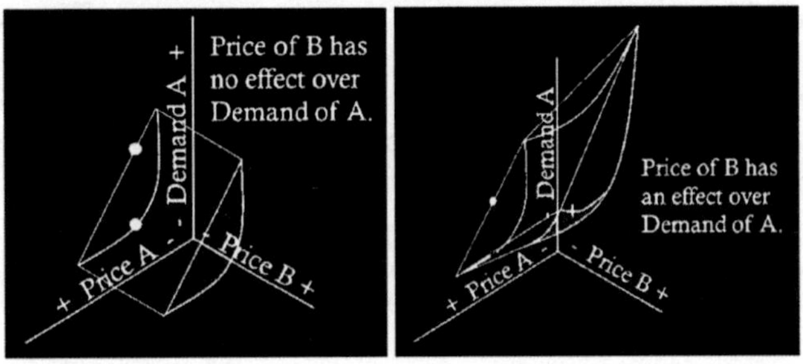


**Fig. 4.** Cross price elasticity. Price change on product B impacting demand of Product A.

Depending on the sign of the cross-elasticity coefficient we can identify whether the product B is complemental or supplemental in relationship with incumbent product A. We can explain this difference using this example:

- Positive (+) Cross-Price Elasticity identifies a Substitute product driving a *Cannibalization effect*. (i.e., *Butter vs margarine*).
- Negative (−) Cross-Price Elasticity identifies a Complement product driving a *Halo-effect*. (i.e., *Peanut butter vs jelly*).

With this cross-price elasticity relationship in mind we can easily expand this concept to a model where we can calculate the demand response of incumbent product A considering 2 or more related products including the original own-price elasticity and even for more products on the numerical example below. On the attached numerical example here below, we have 3 products. Product 1 is the incumbent product and the other two products (2 and 3) these are related products with its current prices definitions as follows.

$$\text{Current.Price}_1 = \$\ 1.45$$
$$\text{Current.Price}_2 = \$\ 1.50$$
$$\text{Current.Price}_3 = \$\ 5.99.$$

Now we can introduce the underlying math required to calculate the total demand response expected for product 1 as a result of all relevant defined interactions from the 3 related products. The first term on the equation is the current or initial demand of the incumbent product 1 (i.e., 893units). As expected, on this example we can verify that we have a specific price elasticity coefficient for each relationship including the own price of the incumbent product. For each relationship we have the magnitude of the price change (i.e., new price versus current price) and also the related price elasticities (i.e., −59.3, −149.9, +13.2). Indeed, depending on supplemental or complemental relationships we can identify positive or negative coefficients. According with this simple math we now calculate the expected demand response for different combinations of prices over products 1, 2, and 3. This is presented on Table 2 below.

$$\begin{aligned} Dem_1 =& 893 - 59.3 * (New.Price_1 - Current.Price_1) \\ &- 149.9 * (New.Price_2 - Current.Price_2) \\ &+ 13.2 * (New.Price_3 - Current.Price_3) \end{aligned}$$

**Table 2.** Cross price elasticity (2) impacting the demand of Product 1.

| New.Price 1 | New.Price 2 | New.Price 3 | New.Dem 1 |
|---|---|---|---|
| $1.45 | $1.20 | $4.78 | 922 |
| $1.45 | $1.35 | $5.37 | 907 |
| $1.45 | $1.50 | $5.97 | 893 |
| $1.45 | $1.65 | $6.57 | 878 |
| $1.45 | $1.80 | $7.16 | 863 |

On the **predictive modeling** the predictor variables reveal complex relationships at product-store level including actual prices (i.e., used for own elasticity and cross-effects), promotional discounts, seasonality, holidays, special events and out of stocks. Indeed, these price and promotion decision variables are interacting with other explanation variables when seeking to predict the demand-response. On supervised machine learning field there is two important techniques: Parametric and Non-Parametric models. The potential disadvantage of a parametric approach is that the model chosen will usually not match the true unknown form of the demand response function. On the other hand, Non-Parametric models are great to handle non-linearities using techniques like Regression-Trees (i.e., XGBoost), Support Vector Machines (SVMs) and Neural Networks. A collection of different forecasts using parametric and non-parametric techniques is used as the proposed approach. Multivariate Ridge regression Log-model is used to get the underlying coefficients to estimate both: (a) own-price elasticities and also (b) the cross-price effects when used for related products (i.e., complement and substitutable products). Cannibalization arise when we have price interactions among substitutable products. On this context the Cannibalization indicates when the demand of a product depends not just on the price of that product but also on the price of substitutable products. The other case is "Halo-effect" which arise when we have price interactions among complemental products. We have positive cross price elasticities for complemental products and negative cross price elasticities for substitutable products.

Here below on attached Table 3 we are presenting the summary results obtained using Ridge regression models grouped by "product type". From left to right we are including for each relevant group the number or products and the average number of weeks included on the historical data. Then we are calculating a forecast error metric (i.e., MAPE or mean absolute percentage error) as a measure of accuracy of the underlying predictive models. On this metric the lower the error value the best is the accuracy of the generated forecast. The column is the variance explained (R2) and the power of the predictive model (F Calc). The next column is a density metric which is measuring the number of elasticities coefficients generated in average for each incumbent product on each different group. As expected, the larger the number means that the predictive modeling is including more interactions on the network. We can verify from Table 3

**Table 3.** Ridge predictive modeling results: accuracy vs density vs explained variance.

| Product type | #of Recs | MAPE error | R2 | F calc | Density (# of elast) | Corr density vs R2 |
|---|---|---|---|---|---|---|
| G1 | 96.35 | 0.11 | 0.44 | 20.42 | 6.83 | 0.47 |
| G2 | 86.70 | 0.16 | 0.36 | 28.18 | 5.96 | 0.31 |
| G3 | 93.24 | 0.12 | 0.38 | U.79 | 6.92 | 0.33 |
| G4 | 81.16 | 0.16 | 0.44 | 18.10 | 6.91 | 0.37 |
| G5 | 93.00 | 0.16 | 0.37 | 19.71 | 5.84 | 0.44 |
| G6 | 93.59 | 0.18 | 0.32 | 21.83 | 4.67 | 0.45 |
| G7 | 68.60 | 0.19 | 0.49 | 424.44 | 4.42 | 0.29 |
| G8 | 91.34 | 0.27 | 0.27 | 11.33 | 2.81 | 0.60 |
| G9 | 86.01 | 0.27 | 0.29 | 14.34 | 4.41 | 0.40 |
| G10 | 92.63 | 0.24 | 0.27 | 15.00 | 4.14 | 0.31 |
| G11 | 94.38 | 0.25 | 0.30 | 15.61 | 3.02 | 0.62 |
| G12 | 102.10 | 0.16 | 0.33 | 16.05 | 5.93 | 0.44 |
| G13 | 96.78 | 0.07 | 0.41 | 19.81 | 7.84 | 0.31 |
| G14 | 71.48 | 0.22 | 0.42 | 42.92 | 5.89 | 0.14 |
| G15 | 86.45 | 0.26 | 0.25 | 11.82 | 4.02 | 0.45 |
| G16 | 88.91 | 0.24 | 0.33 | 21.81 | 4.56 | 0.51 |
| G17 | 57.87 | 0.37 | 0.24 | 24.22 | 1.69 | 0.48 |
| G18 | 91.11 | 0.25 | 0.18 | 18.47 | 0.92 | 0.16 |
| G19 | 81.36 | 0.18 | 0.36 | 11.36 | 7.49 | 0.27 |
| G20 | 69.86 | 0.32 | 0.37 | 25.94 | 3.96 | 0.35 |
| G21 | 104.31 | 0.08 | 0.44 | 25.10 | 7.23 | 0.31 |
| G22 | 90.30 | 0.19 | 0.42 | 21.50 | 6.12 | 0.34 |
| G23 | 86.32 | 0.12 | 0.46 | 15.10 | 8.59 | 0.15 |
| Grand total | 87.06 | 0.20 | 0.35 | 36.22 | 5.29 | 0.37 |

that in the overall summary we are getting in average 5.3 elasticities for each incumbent product. Finally, on the right of the table we are calculating the Pearson correlation of coefficients density versus explained variance (R2). These results are consistent and aligned to what would be the expected pattern.

Besides the Ridge-regression model introduced on the previous paragraph, this paper is proposing alternative forecast models relying on 4 different machine learning techniques including Elastic-Net, XGBoost, Kernel Support Vector Machine and Bayes Regression. For Ridge and Elastic Net as parametric models, the approach includes regularization using L1 and L2 parameters to mitigate overfitting and multicollinearity issues. For the other 3 non-parametric models the approach is to include a hyperparameter optimization to increase the accuracy of the model. Hyperparameters like learning rate (alpha) and maximum no. of iterations were tuned using GridSearchCV technique and applying a 5 fold cross-validation approach. The learning rate of 0.01 and the maximum number of iterations of 1000 were chosen as a result of the experimentation.

The final forecast and elasticities are generated using these 5 different forecast regressors as input vectors into an advanced Ensemble meta-model. The ensemble modeling is an advanced predictive technique seeking to improve the accuracy and confidence on the estimators according with all the available prediction vectors included on the modeling. Its power is reflected on how the ensemble model can capture and include the best of different model techniques which are doing a better job depending on the interaction of the underlying independent variables. An important step of the ensemble modeling is the price elasticity coefficients generation. This component is intrinsic related to what is the concept of model Selection. Model selection is the process where we are comparing the variance explained by the specific predictive model versus the number of price elasticity coefficients included on the model (i.e., a ratio metric analogous to F calculation). Indeed, the proposed approach is the remove from the F calculation the other non-related price-elasticity coefficients. We can see on Fig. 5 that the explained and non-explained variance is relying on the price elasticities factors only and removing the rest of the factors (i.e., sum of squares of the model and residuals). Accordingly, when applying degree of freedom of the factors to calculate the mean squares we are including just the counting of the price-elasticity coefficients.

In general, for ensemble modeling there are 2 approaches: weighted and conditional. These 2 variants are explained on Fig. 6. This ensemble approach was chosen with the consideration in mind that each of the individual regressors are not 100% able to learn all aspects of the nature of the demand response. Thus, the ensemble approach was able to capture all aspects of the data and delivered positive results. Rather than use the average or median of the regressors as final prediction, our ensemble meta-model utilized the last 8 weeks of historical data (i.e., a defined recency parameter) in order to estimate the optimal weights for the underlying voting system. The ultimate goal of the ensemble meta-model is to estimate a better forecast combining all these input vectors. All the vectors included on the ensemble meta-model are trained using the same historical data and processed at the most granular level of detail for every product-store combination.

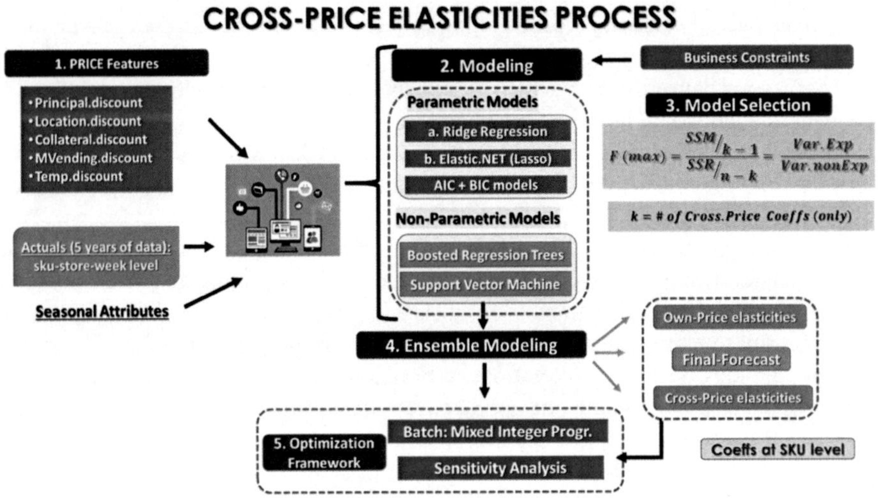


$$F\,(max) = \frac{SSM/_{k-1}}{SSR/_{n-k}} = \frac{Var.Exp}{Var.nonExp}$$

**Fig. 5.** Model selection and price elasticities generation.

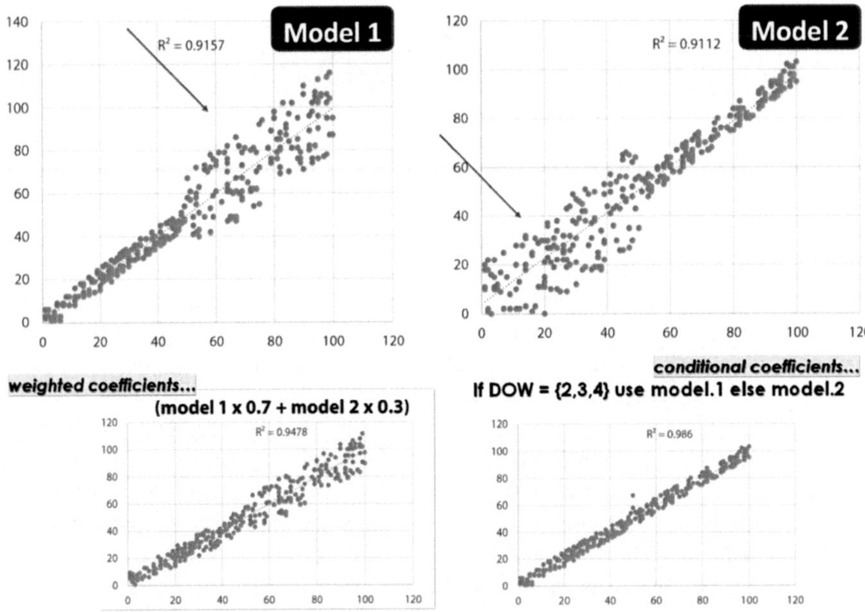


**Fig. 6.** Ensemble modeling: weighted versus conditional approaches.

In our implementation We applied a conditional approach as a conservative strategy to regularize and calculate the absolute smallest elasticity for each interaction. Specifically, for each interaction if the sign of Ridge and Lasso elasticities are the same (i.e., aligned) then we just took the coefficient with the smallest absolute value (i.e., either positive or negative). We can report that on the overall setting we found that around 72% of the interactions (i.e., elasticities) Ridge and Lasso have the same sign on the coefficients. Otherwise, if the sign of Ridge and Lasso elasticities are different then we do an average calculation of these two. For the sake of the extension of the present paper the details about the predictive modeling implementation and how the fitness process was designed during the meta-model final ensemble technique will be presented on a future report.

## 6 Optimization Modeling

As discussed on Sect. 3, existing studies employing MIP strategies have strong restrictions and fail in considering cross-price effects because the computational cost exponentially increases when increasing the number of product interactions. This is especially important and difficult when dealing with real-world large instances with millions of products and stores combinations. This is exactly where the paper approach is seeking to fill the gap. In general, the solution approach is based on approximating and relaxing the model by exploiting the discrete nature of the problem relying on a finite set of relevant price-points. In order to do so, this paper implemented a correlation rule to select the cross-price effects from related products that are the "most influential" in predicting the demand of any given incumbent product. Parametrically speaking, we can control the number of cross interactions, related products and coefficients matrix density that is generated on all the predictive models (i.e., parametric & non-parametric). More importantly, using this strategy to control the cross-price coefficients density on the output matrix then dimensionality and complexity can be mitigated and relaxed on the optimization MIP modeling. In simple terms this simplification allows to decompose and simplify the final optimization problem as we do not need to deal with all the products simultaneously. Indeed, we can decompose the original whole problem into a sequence of un-related simpler problems. This approach drastically reduces the dimensionality of the solution space and therefore we are on a better situation to find the global optimal solution more easily. With own price elasticity and selected/relevant cross-price elasticities in place, we get a reduced finite set "k" of discrete price points for each individual product and the corresponding demand response.

An advanced **optimization model** is proposed relying on the cross-price elasticities coefficients generated from the previous predictive models. This optimization problem could be defined as a concave binary quadratic problem (BQP). The structure of this simplified problem still has an intrinsically stochastic response in the target demand variable regardless on how we define the price-points on the decision variables. This is true whether the prices are defined as continuous or discrete decision variables. Either way, BQP problems are NP-hard and computationally intractable. Moreover, we are dealing with parametric and non-parametric predictive models for the demand response. Therefore, we transformed all outputs from the predictive modeling into a discrete mathematical representation (i.e., an ensembled forecast for the demand response). The resulting model is defined as a Quadratic Mixed Integer problem (QMIP) but it is still no-linear on the objective function as we have a discrete set of price-points and demand-responses required to maximize the revenue. Therefore, the problem has a non-convex representation and it is still NP-Hard. As a result, the problem was again reformulated into a linear Mixed Integer problem (MIP) using McCormick-envelope transformations.

Finally on the proposed modeling a set of **linear constraints** are introduced to handle different type of business rules expressed as regular linear constraints.

a. Business rules for each individual product:

- Demand and prices should be positive constrained relying on historical observed data.
- Prices are chosen from a discrete price ladder. For each product, there is a finite set of permissible prices defined by a binary decision variable. Indeed, the model must consider a bounded solution space to stay away from extreme prices.
- Lower and Upper bounds for prices. A min/max % change guardrail in the pricing solution space. This is a user defined parameter. This is to prevent drastic changes in the proposed discounts.
- Limited number of discounts. The retailer requires to limit the discount frequency for a product (i.e., prevent deal-seekers).

b. Jointly pricing rules for a set of products (i.e., cross-price business rules):

- Cross-effects sign constraints: products in the same category by definition are substitutes and their cross-price elasticities should be constrained as positive. Complemental products have no sign constraints.
- Size constraints. Smaller size of products should have a lower price than larger-sized products. In fact, some of the cross elasticities effects may be incorporated indirectly by restricting the range of feasible prices on the discrete ladder based on informed judgement.

## 6.1 MINLP (Mixed Integer Non-linear Programming) Formulation

**Sets:**

$P = setsofproducts = \{1 \ldots n\}$
$H = setofpricepoints = \{1 \ldots h\}.$

**Parameters:**

$D_i$ = actual demand for product $i$; $i \in P$.
$P_i$ = actual price for product $i$; $i \in P$.
$LB_i$, $UB_i$ = lower & upper for optimal demand for product $i$ (% measured from actual demand); $i \in P$.
$PP_{ij}$ = price point j defined for product $i$; $i \in P, j \in H$.
$PLB_i$, $PUB_i$ = lower & upper for optimal prices for product $i$ (both can be derived from $PP_{ij}$); $i \in P$.
$C_i$ = unit cost for product $i$; $i \in P$.
$M_i$ = profit margin for product $i$; $i \in P$.
$CM_i$ = Competitor reference price for product $i$; $i \in P$.
$ELT_{ik}$ = crosss price elasticitiy of product k when interacting with incumbent product $i$; $i, k \in P$.

**Variables:**

$NP_i$ = optimal price defined for product $i$; $i \in P$.
$ND_i$ = predicted demand for product $i$; $i \in P$.

$$XP_{i,j} = \begin{cases} 1 \text{ if price point } j \text{ is defined as optimal for product } i;\ i \in P, j \in H \\ 0 \text{ otherwise} \end{cases}$$

$$HP_i = \begin{cases} 1 \text{ if optimal price defined for product } i \text{ is higher than competitor};\ i \in P \\ 0 \text{ otherwise} \end{cases}$$

**Objective function:**

$$Maximize\, Z = \sum\nolimits_{i \in p} NP_i \times ND_i \quad (1)$$

**Subject to:**

$$ND_i, NP_i \geq 0; \forall i \in P \quad (2)$$

$$\sum\nolimits_{j \in H} XP_{ij} = 1; \forall i \in P \quad (3)$$

$$NP_i = \sum\nolimits_{j \in H} XP_{ij} * PP_{ij}; \forall i \in P \quad (4)$$

$$NP_i \leq (1 + M_i) * C_i; \forall i \in P \quad (5)$$

$$NP_i = NP_j; \forall R_i = R_j i, j \in P \quad (6)$$

$$NP_i \leq CM_i * (1 + HP_i); \forall i \in P \tag{7}$$

$$\sum\nolimits_{i \in p} HP_i \leq PRCH \tag{8}$$

$$(1 - LB_i) * D_i \leq ND_i \leq (1 + UB_i) * D_i; \forall i \in P \tag{9}$$

**Cross-Price Effects (Cannibalization)**
The following section discusses a formulation that considers the cross-price effects from a subset of related products. The cross-price elasticities represent the demand response of an incumbent product to the prices of other related products. This motivates considering the subset of related products when predicting the demand for complimentary or supplementary products at different price-points. It is important to understand the individual contributing effects. We can derive these individual impacts from Eq. (10).

$$ND_i = D_i + \sum\nolimits_{k \in p} ELT_{ik} * (NP_k - P_k); \forall i \in P \tag{10}$$

We can see from Eq. (10) that optimization model is executed as a network setting considering all related products simultaneously rather than solving the problem sequentially for isolated individual products. The individual contributions and revenue lifts coming from related products are computed prior to the optimization according with the cross-price elasticities coefficients generated over the underlying predictive modeling.

## 6.2 MILP (Mixed Integer Linear Programming) Formulation

It was mentioned in Sect. 4 our problem has a non-linear objective function on the revenue as we are multiplying prices and demands to be maximized on a quadratic function. This paper proposes to address this challenge including McCormick-envelope transformations in order to be able to solve large scale instances in realistic time. McCormick Envelopes techniques are based on convex relaxations used in Non Linear Programming problems. In general, solving non convex problems is difficult as the non convex function is transformed into a convex function by relaxing the underlying parameters on the problem and decreasing the computational difficulty but at the cost of expanding the solution space and adding potential solutions that do not necessarily corresponds to the original objective function. However, solving the relaxed convex problem will provide a lower bound for the optimal solution and mitigate the challenge to deal with local minima solutions that the algorithm may interpret as the global solution. Therefore, the tricky part is to identify a convex relaxation that has the tightest possible bounds. McCormick Envelope is one particular kind of relaxation that guarantees convexity but keeps the bounds sufficiently tight as they retain convexity while minimizing the size of the new feasible region [28]. Solving the relaxed convex problem will identify the lower bound solution for the original problem. An upper bound can be determined by solving the original non convex problem using the values obtained from the relaxed problem and then checking for feasibility. From quadratic Eq. (1) we transform it using McCormick Envelopes into this equivalence as follows:

$$Minimize\, Z = \sum\nolimits_{i \in p} -W_i \tag{11}$$

$$W_i \geq (1 - PLB_i) * P_i * ND_i + NP_i * (1 - LB_i) * D_i - (1 - PLB_i) * P_i * (1 - LB_i) * D_i \tag{12}$$

$$\forall i \in P$$

$$W_i \geq (1 + PUB_i) * P_i * ND_i + NP_i * (1 + UB_i) * D_i - (1 + PUB_i) * P_i * (1 + UB_i) * D_i \tag{13}$$

$$\forall i \in P$$

$$W_i \leq (1 + PUB_i) * P_i * ND_i + NP_i * (1 - LB_i) * D_i - (1 + PUB_i) * P_i * (1 - LB_i) * D_i \tag{14}$$

$$\forall i \in P$$

$$W_i \leq NP_i * (1 + UB_i) * D_i + (1 - PLB_i) * P_i * ND_i - (1 - PLB_i) * P_i * (1 + UB_i) * D_i \tag{15}$$

$$\forall i \in P$$

Inequalities (12–15) are valid envelop relaxation constraints as $P_i$ and $D_i$ are model parameters and all of these have defined lower and upper bounds.

## 7 Results and Discussion

On this section we will discuss summary and results at high level. On attached Table 4 from left to right we are reporting high level results grouped by Product type for the Ridge results model. For each group we are including the number of significant relevant product interactions on the assortment. For Ridge results we have the average number of weeks we are consuming for the historical data, also the variance explained (i.e., R2) and the predictive power test indicator (i.e., Fcalc) for each group. The next column Avg. No.clast corrcsponds to the average number of interactions we are capturing for each incumbent product. On this regard, we can verify that on overall across all groups we are getting 8.3 cross-price elasticities. This number means that we are dealing with a very dense and rich network of interactions which in turn will increase the solution space complexity on the optimization modeling. The other 2 columns Avg (+) and Avg (−) indicate how the elasticities are divided by positive or negative effects correspondingly. Lift units (+) and Lift units (−) are the corresponding units Lifts according with positive and negative elasticities. With these 2 metrics we can have an insight about how important the overall effects at category level for are supplemental (−) versus complemental (+) effects. We can use these 2 to calculate a Ridge Lift expressed in percentage. Finally on the Ridge block we include a Coeffs Var metric to estimate the variance of the elasticities across all the identified interactions.

**Table 4.** Predictive modeling results: cross-price elasticities summary (Ridge model).

| Product type | Avg nrec | Avg R2 | Avg Fcalc | Avg No.elast | Avg (+) | Avg (−) | Ridge lift % | Coeffs var |
|---|---|---|---|---|---|---|---|---|
| Gl | 85 | 28% | 4.9 | 8.2 | 3.7 | 4.4 | −10.0% | 114.2 |
| G2 | 74 | 32% | 4.8 | 8.1 | 3.7 | 4.3 | −5.0% | 29.5 |
| G3 | 88 | 25% | 3.8 | 8.0 | 3.6 | 4.2 | 1.4% | 26.9 |
| G4 | 78 | 38% | 5.7 | 9.2 | 4.1 | 5.0 | −11.0% | 118.3 |
| GS | 78 | 16% | 1.7 | 7.4 | 3.3 | 3.3 | 3.1% | 2.7 |
| G6 | 82 | 32% | 8.1 | 8.3 | 3.9 | 4.3 | −9.0% | 92 |
| G7 | 45 | 2% | 10.1 | 0.2 | 0.0 | 0.2 | −16.0% | 5.2 |
| G8 | 85 | 23% | 2.5 | 8.3 | 3.6 | 4.5 | −10.5% | 28.7 |
| G9 | 85 | 29% | 4 | 8.7 | 4.2 | 4.3 | −5.9% | 12.9 |
| G10 | 56 | 26% | 2.7 | 7.0 | 4.0 | 2.7 | 24.3% | 12.3 |
| G11 | 97 | 24% | 2.6 | 9.1 | 4.4 | 4.6 | −1.7% | 43 |
| G12 | 82 | 24% | 2.7 | 7.3 | 3.2 | 3.8 | −8.3% | 6.6 |
| G13 | 79 | 21% | 2.2 | 8.4 | 3.7 | 4.4 | −4.6% | 9.9 |
| G14 | 76 | 38% | 5.9 | 9.2 | 4.4 | 4.7 | −1.6% | 7.6 |
| G15 | 78 | 26% | 19.5 | 7.8 | 3.9 | 3.5 | 9.0% | 56.4 |
| G16 | 66 | 24% | 4.2 | 5.7 | 3.0 | 2.5 | 15.3% | 12 |
| G17 | 88 | 25% | 3 | 8.8 | 4.2 | 4.5 | −5.3% | 62.9 |
| G18 | 93 | 28% | 5.7 | 8.7 | 4.5 | 4.2 | 0.4% | 90.7 |
| G19 | 70 | 29% | 3.1 | 9.0 | 4.1 | 4.6 | −1.9% | 17.8 |
| G20 | 71 | 24% | 2.6 | 7.8 | 3.9 | 3.5 | 2.8% | 15.1 |
| G2l | 71 | 24% | 2.3 | 7.9 | 4.4 | 3.1 | 19.9% | 11.7 |
| G22 | 84 | 27% | 4.5 | 8.0 | 4.4 | 3.3 | 15.4% | 7.4 |
| G23 | 75 | 12% | 1.5 | 7.0 | 3.5 | 2.6 | 14.1% | 5.4 |
| Grand total | 76 | 26% | 3.5 | 8.3 | 3.9 | 4.1 | −1.2% | 37.6 |

The same indicators reported for the Ridge modeling block are then replicated for the Lasso modeling (see Table 5). As expected on Lasso we have a smaller number of elasticity interactions and of course a smaller variance explained when compared with Ridge. However, the predictive power in F Calc is larger. More importantly the Lasso Lift indicators (i.e., units and percentage) and coefficients variance is larger on Lasso. This finding is very important as this is a true indicator about the nervousness or steadiness of the entire system. In other words, Ridge elasticities (i.e., positive or negative) are more conservative as the lift is distributed among a larger set of related products.

**Table 5.** Predictive modeling results: cross-price elasticities summary (Lasso model).

| Product type | Avg nrec | Avg R2 | Avg Fcalc | Avg No.elast | Lasso lift % | Coeffs var |
|---|---|---|---|---|---|---|
| Gl | 85 | 22% | 10.4 | 3.2 | −8.00% | 169.3 |
| G2 | 74 | 30% | 10.6 | 3.3 | −2.50% | 11.8 |
| G3 | 88 | 23% | 8.4 | 2.9 | 2.80% | 54.6 |
| G4 | 78 | 35% | 10.9 | 4.3 | −11.50% | 153.9 |
| GS | 78 | 13% | 4.4 | 1.9 | 4.30% | 0.7 |
| G6 | 82 | 34% | 13 | 5.4 | −11.30% | 295 |
| G7 | 45 | 0% | 0.7 | 0 | −4 | 1.3 |
| G8 | 85 | 17% | 7.3 | 2.6 | −6.80% | 21 |
| G9 | 85 | 26% | 9.3 | 3.3 | −0.70% | 4.5 |
| G10 | 56 | 25% | 4.8 | 3.4 | 21.50% | 11.7 |
| G11 | 97 | 22% | 6.7 | 3.5 | −0.10% | 40.1 |
| G12 | 82 | 19% | 6.4 | 2.4 | −6.20% | 1.6 |
| G13 | 79 | 16% | 6.6 | 2 | 1.60% | 2.9 |
| G14 | 76 | 40% | 11.4 | 5 | −2.30% | 9.7 |
| G15 | 78 | 24% | 65.3 | 3.1 | 2.00% | 51.2 |
| G16 | 66 | 24% | 7 | 3 | 10.30% | 4.2 |
| G17 | 88 | 17% | 7.3 | 2.5 | −3.10% | 39.4 |
| G18 | 93 | 24% | 12.6 | 2.8 | 16.20% | 72.4 |
| G19 | 70 | 24% | 8.5 | 2.6 | 1.40% | 5 |
| G20 | 71 | 12% | 5.7 | 1.4 | 6.30% | 7.8 |
| G2l | 71 | 17% | 6.4 | 2.1 | 23.50% | 6.4 |
| G22 | 84 | 23% | 9.2 | 2.6 | 19.40% | 47.2 |
| G23 | 75 | 7% | 5.2 | 1.2 | 6.80% | 19.3 |
| Grand total | 76 | 21% | 8.5 | 2.6 | 1.60% | 59.3 |

Below on Table 6 we are reporting the summary results for the optimization models. For optimization modeling we are relying on the price elasticities generated using the ensemble technique. Indeed, we are combining previous Ridge & Lasso elasticities presented on Table 4 and these are combined/ensembled to generate the final elasticities coefficients to be used on the optimization modeling. As presented on Sect. 5 there are weighted and conditional approaches for ensemble modeling. We applied a conditional approach as a conservative strategy to regularize and calculate the absolute smallest elasticity for each interaction. Specifically, for each interaction if the sign of Ridge and Lasso elasticities are the same (i.e., aligned) then we just took the coefficient with the smallest absolute value (i.e., either positive or negative). We can report that on the overall setting we found that around 72% of the interactions (i.e., elasticities) Ridge and Lasso have the same sign on the coefficients. Otherwise, if the sign of Ridge and Lasso elasticities are different then we do an average calculation of these two. Because of confidentiality reasons we are just reporting increase in demands and revenues as percentages. Then we include the number of products with price markdowns and markups and also the average percentage of price changes defined as optimal for each group. With these new optimal price changes defined for each individual product we can easily calculate the new Optimal demand response and new Optimal revenue reported on each individual group of products. Interestingly from Table 6 we can verify that on the overall setting we have that for around 60% of the products we are decreasing the price of the products on the optimal solution.

**Table 6.** Prescriptive modeling results: optimal prices, demand response & revenue lift.

| Product type | Price (−) | Price (+) | Avg Price change | % (+) Demand | % (+) Revenue |
|---|---|---|---|---|---|
| Gl | 13 | 8 | 0.99 | 10.20% | 9.40% |
| G2 | 22 | 19 | 1 | 4.20% | 3.20% |
| G3 | 18 | 18 | 1.02 | 6.10% | 5.10% |
| G4 | 8 | 6 | 0.99 | 10.30% | 9.90% |
| G5 | 17 | 11 | 0.97 | 8.40% | 14.00% |
| G6 | 5 | 4 | 0.99 | 10.00% | 8.50% |
| G7 | 11 | 8 | 0.99 | 9.20% | 9.40% |
| G8 | 12 | 6 | 0.96 | 6.90% | 5.60% |
| G9 | 13 | 8 | 0.97 | 8.00% | 7.40% |
| G10 | 7 | 8 | 1.03 | 10.00% | 10.10% |
| G11 | 4 | 2 | 0.97 | 9.80% | 11.30% |
| G12 | 38 | 11 | 0.93 | 7.80% | 6.50% |
| G13 | 12 | 4 | 0.94 | 5.40% | 5.20% |
| G14 | 3 | 3 | 1 | 7.50% | 7.40% |
| G15 | 2 | 3 | 1.01 | 10.20% | 5.60% |
| G16 | 13 | 12 | 1.01 | 8.10% | 6.60% |
| G17 | 4 | 3 | 1.02 | 8.80% | 6.70% |
| G18 | 47 | 24 | 1 | 3.80% | 2.50% |
| G19 | 101 | 62 | 0.96 | 6.30% | 6.40% |
| G20 | 32 | 22 | 0.97 | 7.50% | 6.80% |
| G21 | 9 | 8 | 1 | 6.40% | 5.70% |
| G22 | 5 | 4 | 0.97 | 2.60% | 4.50% |
| G23 | 35 | 26 | 0.99 | 4.10% | 2.80% |
| Grand total | 534 | 349 | 0.98 | 6.30% | 5.70% |

# 8 Business implementation and value generation

The goal on this section is to present the results obtained during the test and deployment of the proposed solution on a real-world business context. The solution provides the complete pricing recommendations and corresponding sales and revenue predictions for all products and stores during a defined time-period. There are 3 business topics included here: operationalization, organizational acceptance and business value.

## 8.1 Operationalization

The process described on the present paper was designed to optimize pricing decisions on a weekly basis. We were able to model the historical customer demand on weekly cycles across all detailed product-store combinations re-training the underlying demand-based models. Accordingly, price elasticity effects can be changing on a weekly basis as well. Important efforts across different areas of the company were required in order to execute the proposed price strategy and monetize the opportunity at the stores. This is true not only because of the efforts required on the stores to change the prices more dynamically but also because of the rules and constraints that were implemented to make the operationalization of the solution more consumer-friendly (i.e. rounding, competitor price bounds, etc.). It is important to consider the retailer under scope on this work does not have an on-line capability to change the prices. Indeed, implement a price change takes a more complex action manually changing the labels seen by the customers at the store. Without consideration of these features the solution would be poorly integrated and hard to manage. To our best knowledge, this is the first a controlled scientific experiment is being implemented using this kind of complexities.

## 8.2 Adoption/Organizational Acceptance

One of the most important challenges we faced in our implementation on the business process was to stay-away from the "black box" effect. Indeed, price recommendations were accompanied with the underlying generated cross-price elasticities coefficients used to replicate the forecasts. This approach facilitates the understanding of why a price change is being recommended addressing any concerns about user's mistrust. It also enables what-if scenario analysis relative to a baseline of specified pricing decisions as well as the visualization of the expected revenue and profit corresponding to different possible price-points in a given situation. Interestingly, a cultural impact resulted from the introduction of the proposed methodology. It changed retailer's approach about the pricing activity from an intuition-based to a more model-based method. This demonstrated that pricing decisions can be improved by a scientific approach. Interesting, the model solution created a consensus and a basis for manager's discussion providing stronger arguments to justify their intuition. At the beginning of the project, the forecast received most of the attention in the meetings and the process was evaluated based on the forecast accuracy. It took a lot of effort and management change to shift the discussion to the impact of the suggested prices on revenue. This is considered a key learning and a milestone because many price optimization projects do not monetize the benefits and finish just with forecast accuracy indicators.

### 8.3 Business Value and Actionable Insights

The financial impact of the solution provided is explained by the model's ability to maximize the revenue and maintain control of the profit margin at the same time. Furthermore, the assortment has been optimized in terms of complexity and business value. Given the successful results, other category managers have shown interest in adapting this solution for their own categories. Our solution provided actionable insights on what's happening, why it's happening and what we can do to get better results. The ability to model cross price elasticities is the core of any price optimization solution as we unlock the ability to estimate the effect that a change in the price have on the demand response. The success of the proposed price optimization approach was not just about a tool to increase the revenue and the profit margin, but it was also about how the company can deal with category and assortment complexities at operation level at stores and also at tactic level with vendors. Indeed, the insights generated here for the assortment are extremely valuable when having conversations with vendors about campaigns and product shelf-space allocation.

## 9 Conclusion and Future Challenges

We addressed on this paper one of the biggest challenges in retailing which is setting an optimal price for products seeking to maximize the generated revenue for the business while taking control of profit margin and assortment complexity on the category. The optimal price is the middle ground of 2 objectives between the customer demand response and the marginal profit generated on the business. Data-science models are powerful tools to address this kind of problems as these can continually learn patterns from data with no explicitly re-programming. The present paper contains technical details of the methodology developed and provides a discussion of its implementation. To the best of our knowledge, our paper is the first documented application of a complete price optimization solution implemented as part of a controlled experiment. As part of the methodology a combination of predictive and prescriptive analytics were presented which allow an end-to-end solution with no required human input. The approach presented has two main-components: (P1) a demand response predictive modeling with a relevant cross-price elasticities output; and (P2) cross-price elasticities are used as input into a price optimization model to determine the optimal set of prices required to maximize the revenue and profit margin subject to a large set of assortment and capacity constraints. Cross-price elasticity estimates the demand response at different prices discounts.

**Main contributions can be summarized as follows**

- For P1 a hybrid ensembled demand response model was designed relying on a relevant and finite set of price-points. A significant challenge here was to model cannibalization and halo effects. This challenge is attached to the business setting to dealt with a large set of several thousands of products.

- An efficient prescriptive model was presented for the Price Optimization problem. This paper proposes an advanced MIP model including McCormick-envelope transformations to solve large scale price optimization instances in realistic time. The optimality results were tested relying on the bounds obtained from the MIP formulation computed efficiently using an open source solver with a defined processing time limit.
- Our detailed empirical evaluation presented on Sect. 8 revealed positive results. Predictive ensemble modeling add up an improvement by reducing the baseline error around 10%. On the Optimization side our results presented improving by showing up to a 6% of increase on revenue while controlling profit margin and assortment constraints.

**Learnings and future opportunities**

- Leverage cross-price elasticities modeling into the "category-assortment" optimization problem. Modeling of complex relationships for complementary and substitute products is a hot business topic as nowadays there are large teams managing these complexities in a regular basis.
- In 2020 we saw eCommerce become the norm to deal with Covid-social distancing. The challenge is how we can leverage a price optimization solution to effectively support the potential benefits of the online channels.
- Dynamic pricing on eCommerce using technological enablers like electronic tags, on-line pricing elements, in-store digital messaging and kiosks to improve and speed-up the pricing process on a cheaper and faster way. Digital price tags are enabling brick-and-mortar retailers to do as many price changes as e-commerce sites to fight back more efficiently versus competitors.
- Introduce a more educated process to weight differently the under versus over predicted demands depending on the context of the real cost of lost sales versus the cost of inventory.

Predictive modeling produces two important outcomes: (a) significant cross-price elasticities relationships revealing insights and opportunities for revenue management and category assortment (b) expected demand response when we do experimentation over the underlying price factors. On the optimization side of our solution, we pursued suitability for industry size problems on a realistic scale without oversimplifying the problem. We employed specialized mathematical transformations to recover computational tractability. Prescriptive analytics is recognized as the next step to get most of the business benefit from the massive amount of predictions. Indeed, predictive meets prescriptive analytics when we integrate and leverage the predictive modeling into a mathematical optimization framework.

## References

1. Jacobs, K.: Sears holdings posts unexpected loss on markdowns (2008). http://www.reuters.com

2. Babar, M., Nguyen, P.H., Cuk, V., Kamphuis, I.G.: The development of demand elasticity model for demand response in the retail market environment. In: 2015 IEEE Eindhoven PowerTech, pp. 1–6. IEEE, Eindhoven (2015)
3. Bain & Company. Price optimization models (2015) http://www.bain.com/publications/articles/management-tools-price-optimization-models.aspx
4. Gupta, R., Pathak, C.: A machine learning framework for predicting purchase by online customers based on dynamic pricing. Proc. Comput. Sci. **36**, 599–605 (2014)
5. Kopalle, P.K.: Modeling retail phenomena. J. Retail. **86**(2), 117–124 (2010)
6. Levy, M., Grewal, D., Kopalle, P.K., Hess, J.D.: Emerging trends in retail pricing practice: implications for research. J. Retail. **80**(3), 13–21 (2004)
7. Bertsimas, D., De Boer, S.: Dynamic pricing and inventory control for multiple products. J. Revenue Pric. Manag. **3**(4), 303–319 (2005)
8. Shen, Z.J.M., Su, X.: Customer behavior modeling in revenue management and auctions: a review and new research opportunities. Prod. Oper. Manag. **16**(6), 713–728 (2007)
9. Levin, Y., McGill, J., Nediak, M.: Optimal dynamic pricing of perishable items by a monopolist facing strategic consumers. Prod. Oper. Manag. **19**(1), 40–60 (2010)
10. Ahn, H.S., Gumus, M., Kaminsky, P.: Pricing and manufacturing decisions when demand is a function of prices in multiple periods. Oper. Res. **55**(6), 1039–1057 (2007)
11. Smith, S.: Clearance pricing in retail chains. In: Agrawal, N., Smith, S. (eds.) Retail Supply Chain Management, vol. 223, pp. 387–408. Springer, New York (2009). https://doi.org/10.1007/978-1-4899-7562-1_14
12. Valkov, T.: From theory to practice: real-world applications of scientific pricing across different industries. J. Revenue and Pric. Manag. **5**(2), 143–151 (2006)
13. Talluri, K.T., Van Ryzin, G.J.: The Theory and Practice of Revenue Management, vol. 68. Springer, New York (2006)
14. Ozer, O., Phillips, R.: The Oxford Handbook of Pricing Management. Oxford University Press, Oxford (2012)
15. Bitran, G., Caldentey, R.: An overview of pricing models for revenue management. Manuf. Serv. Oper. Manag. **5**(3), 203–229 (2003)
16. Maglaras, C., Meissner, J.: Dynamic pricing strategies for multi-product revenue management problems. Working paper, Columbia University(2004). http://www.meiss.com
17. Caro, F.: Gallien, J: Clearance pricing optimization for a fast-fashion retailer. Oper. Res. **60**(6), 1404–1422 (2012)
18. Koushik, D., Higbie, J.A., Eister, C.: Retail price optimization at intercontinental hotels group. Interfaces **42**(1), 45–57 (2012)
19. Lee, S.: Study of demand models and price optimization performance. Ph.D. thesis, Georgia Institute of Technology (2011).
20. Ito, S., Fujimaki, R.: Large-scale price optimization via network flow. In: Lee, D., Sugiyama, M., Luxburg, U., Guyon, I., Garnett, R. (eds.) Advances in Neural Information Processing Systems, pp. 3855–3863. Curran Associates Inc, Red Hook (2016)
21. Yabe, A., Ito, S., Fujimaki, R.: Robust quadratic programming for price optimization. In: Sierra, C. (ed.) International Joint Conference on Artificial Intelligence Proceedings. Curran Associates, Inc., Red Hook (2017)
22. Maddah, B., Bish, E.K.: Joint pricing, assortment, and inventory decisions for a retailer's product line. Naval Res. Logist. **54**(3), 315–330 (2007)
23. Cohen, M.C., Kalas, J., Perakis, G.: Optimizing promotions for multiple items in supermarkets. In: Ray, S., Yin, S. (eds.) Channel Strategies and Marketing Mix in a Connected World, vol. 9, pp. 71–98. Springer, Heidelberg (2018). https://doi.org/10.1007/978-3-030-31733-1_4
24. Scikit Learn: Developing scikit-learn estimators (2019). https://scikit-learn.org/stable/developers/develop.html

25. Bienstock, D.: Computational study of a family of mixed-integer quadratic programming problems. Math. Program. **74**(2), 121–140 (1996)
26. Grossmann, I.E.: Review of nonlinear mixed-integer and disjunctive programming techniques. Optim. Eng. **3**(3), 227–252 (2002)
27. Besbes, O., Zeevi, A.: On the (surprising) sufficiency of linear models for dynamic pricing with demand learning. Manag. Sci. **61**(4), 723–739 (2015)
28. Castro, P.: Tightening piecewise McCormick relaxations for bilinear problems. Comput. Chem. Eng. **72**, 300–311 (2015)

# Robo Advisors vs. Value Investing Strategies: A Fuzzy Jensen's Alpha Assessment

Rodrigo Caballero-Fernández[1], Klender Cortez[1](✉), and David Ceballos-Hornero[2]

[1] Facultad de Contaduría Pública y Administración, Universidad Autónoma de Nuevo León, Av. Universidad S/N, Cd. Universitaria, 66451 San Nicolás de los Garza, NL, Mexico
klender.cortezal@uanl.edu.mx
[2] Universitat de Barcelona, 08007 Barcelona, Spain

**Abstract.** Robo-advisor (RA) platforms have become fundamental in the areas of finance and technology, as they have revolutionized the way investments are carried out and managed. Nevertheless, investment strategies using traditional financial management are currently being maintained. The objective of this research is to compare an investment portfolio that utilizes similar strategies to those of a RA against an investment portfolio that makes decisions through a consensus of valuation analysts. Fuzzy Jensen's Alpha is used to compare both portfolios. To create a new portfolio strategy proposal for RA platform as well as a value investing portfolio, we selected Latin American listed companies in Mexico, Colombia, Peru, Chile, and Brazil for the sample. Additionally, we used ETFs to replicate those countries. The timespan considered was from January 15, 2015 to June 28, 2019. The results report that both strategies succeeded in surpassing the benchmark; however, the analysts' portfolio has accelerated its growth since 2018, increasing its positive gap against the RA portfolio. In this sense, the portfolio developed by the analysts gives an average return higher than the average return of the RA. Nonetheless, the RA portfolio has a higher possibility of obtaining abnormal or unexpected returns than the analyst's value investing portfolio, given the systematic risk involved.

**Keywords:** Automated investment strategies · Fintech · Fuzzy regression

## 1 Introduction

Throughout history, technology has been one of the main change factors that has revolutionized the way we work. Technology can make significant changes in every scientific area, including finance. In the financial world, "fintech" (sometimes called fin-tech) is a neologism that takes its origins from the words "finance" and "technology". In the 2000s, before the term "fintech" was popularized, the term "e-finance" was introduced, derived from the use of information and communication technologies in the financial industry [1]. In the fintech field, we can find Robo-Advisor (RA) platforms, which have become a fundamental piece in the financial and technology area, transforming the way


M. d. P. Rodríguez García et al. (Eds.): XX SIGEF 2021, LNNS 384, pp. 178–193, 2022.
https://doi.org/10.1007/978-3-030-94485-8_13

investments are carried out and managed. Thanks to its automated management, RA can minimize commissions and eliminate conflicts of interest [2].

In contrast, exchange-traded funds (ETFs) in the stock market world are a fundamental key to mutual funds in the same period. According to the Investment Company Institute [3], ETFs have developed a huge positioning in markets in the last ten years. Nevertheless, investment strategies with traditional management are currently being maintained. There are endless investment strategies that include factors such as moment, value, and growth. The most popular methodologies are technical analysis and fundamental analysis.

The objective of this research is to compare an investment portfolio that utilizes similar strategies to those of an RA against an investment portfolio that makes decisions through a consensus of valuation analysts. Fuzzy Jensen's alpha is used to compare both portfolios.

RA is recognized as one of the most important disruptive trends in the industry of active management. Terms such as "Robo-Advise", "Robo Advisor" and "Robo Adviser" have become more popular [4]. Starting in 2013, a series of issues have helped RA platforms gain in popularity, including international regulations in favor of investor protections and the tremendous penetration of smartphones in the market. This popularity was reflected in the increase not only in low-profile clients but also in high-capital value clients [5].

In this sense, Bjerknes and Vukovic [6] estimated that the performance of four RA platforms (Betterment, Wealthfront, Schwab Intelligent Portfolios, and Future Advisor) would show results of superior performance to their market standard in terms of accumulated returns during the eight years after the financial crisis. They also concluded that three of the four studied RAs presented a better risk-adjusted return concerning their benchmark. Moreover, Uhl and Rohner [7] carried out a study in which they compared RAs against the traditional way of performing investment consultancies, where their main areas of interest were the commission costs and the costs occasioned by the cognitive biases of humans in traditional advising.

In other studies related to the topic, Ling et al. [8] built an intelligent portfolio management system for RA based on modern portfolio optimization theory. Bird and Whitaker [9] examined a wide selection of value and momentum strategies applied to the major European markets from 1990 to 2002, finding strong evidence that certain implementations of value and momentum investing performed particularly well over this period across the European markets. Furthermore, D'Acunto [10] studied a robot-advising portfolio optimizer that constructs tailored strategies based on investors' holdings and preferences. Adopters are similar to nonadopters in terms of demographics but have more assets under management, trade more, and have a higher risk-adjusted performance. The robo-advising tool has the opposite effects across investors with different levels of diversification before adoption: it increases portfolio diversification and decreases volatility.

Based on previous studies, there are some papers on robo-advising and value investing methodologies, but we did not find a comparative study of these methodologies. In addition, we analyze emerging markets using Latin American stocks to examine AR and compare it with a traditional technique, which has been little analyzed with fintech issues. The principal intention of this research is to compare two investment methodologies. The first one uses financial model theory, specifically, optimizing the Sharpe ratio, as some RAs are usually used in real life. The second methodology uses the value investing technique, which is practiced only by humans.

The paper is divided into six sections, starting with the introduction, after which it continues with the literature review, the methodology, the results, the conclusions, and the references.

## 2 Literature Review

### 2.1 Value Investing

The value investing methodology was started in 1920 by Benjamin Graham and David Dodd [11]. After them, there were many authors who defended this investment methodology. Fama and French [12] used a three-factor model to select stocks expecting a better than average performance. Additionally, it is important to recognize that the analyst who creates the value investing model tends to give much importance to risk management; however, mistakes are tended to be made in the free cash flow forecast [13].

Asness et al. [14] prove that the efficacy of company value is a prime of acquired risk; however, anomalies of the irrational behavior of the market increase the risk of falling prices. In other words, we can say that the value investing methodology is an advanced grade over the decision-making of a portfolio manager and financial areas, but we need to remember that is not an exact science; the valuations can represent the cognitive bias of the analysts who develop the models, overestimating or underestimating the importance of some variables in the operations.

On the other hand, there are some authors who have discovered some problems with this value investing methodology. For example, some experiments have revealed that analysts who make forecast reports of companies in the stock market tend to obtain anchorage bias. Additionally, a study discovered a significant systemic anchoring cognitive bias in the 2- and 10-year treasury bills of the United States of America [15]. Ototeye and Siddiquee [16] analyzed the importance of cognitive bias in the value investing method. They found that the field of psychology suggests that when there is a complex decision with a high level of uncertainty, humans tend to fall into heuristics to simplify the information.

### 2.2 Sharpe Ratio

The main portfolio characteristics used in portfolio optimization suggested by Markowitz [17] are expected return and risk. In contrast to the expected return, the specification of the portfolio risk appears to be a more complicated task. In Markowitz's portfolio theory, the variance is taken as a risk measure. Then, optimal portfolios are constructed by minimizing the variance for a given level of the expected return or by maximizing the expected return for a given value of the variance.

However, these optimization problems consider only one characteristic of the portfolio, i.e., the risk. Another possibility is to maximize the Sharpe ratio (SR), which is defined as a ratio of the expected portfolio return to the standard deviation [18]. The Sharpe ratio, also called the variability ratio, is measured as follows:

$$Sharpe\ ratio = \frac{R_p - R_f}{\sigma_p} \tag{1}$$

where:
$R_p$ is the return of the portfolio,
$R_f$ is the risk-free rate; and
$\sigma_p$ is the standard deviation of the portfolio.

The numerator of Eq. (1) shows the difference between the funds' average annual return and the pure interest rate; it is thus the reward provided to the investor for bearing risk. The denominator measures the standard deviation of the annual rate of return, i.e., it shows the amount of risk taken. The ratio is thus the reward per unit of variability [19].

### 2.3 Robo Advisors

Currently, RA platforms have become stronger in certain countries, as shown in Fig. 1. RA first appeared between 2008 and 2010 as part of the larger fintech phenomenon, which has disrupted the different areas of finance through technology. Since 2013, a series of issues have helped RA platforms gain in popularity, including international regulations in favor of investor protections and the several penetrations of smartphones in the market. This popularity was reflected not only in low-profile clients but also in high capital value clients [20].

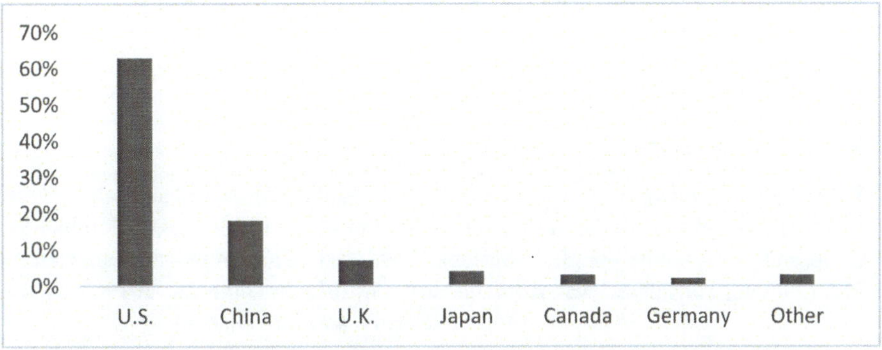


**Fig. 1.** Countries using Robo-Advisor platforms. Source: Own elaboration.

RA platforms always offer passive investment strategies focusing on three main areas: asset allocation, tracking portfolio and portfolio rebalancing. Likewise, there are mainly two types of RA: independent start-ups such as Betterment and Wealthfront and Robo-advisory platforms created by well-positioned investment companies such as Vanguard and Black Rock [21].

RA differs from online platforms in two principal ways: online questionnaires for assigning the client's risk profile and portfolio management, which includes financial instruments that do not require active intervention in decision-making, such as

ETFs. This combination of financial instruments and algorithms can significantly reduce administrative costs through complete automation [22] (Fig. 2).

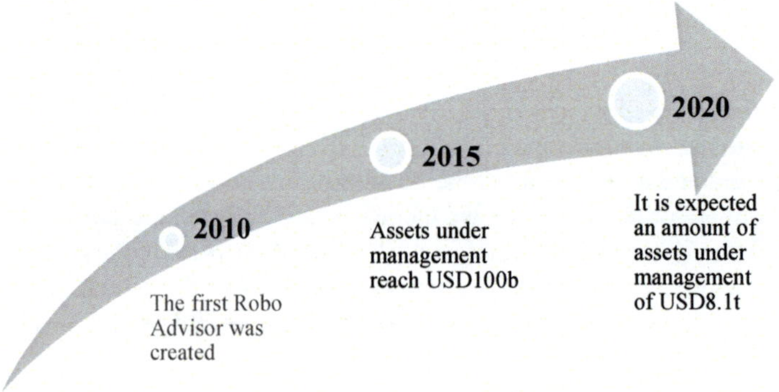


**Fig. 2.** Growth of Robo Advisor portfolio. Source: Own elaboration.

Previous studies have investigated different forms of strategy performances for portfolio management, but we have not found a thorough research study comparing a similar strategy to RA against a strategy developed by a consensus of firm value advisors. For the above, we formulate the following hypothesis: *Automated investment strategies, similar to those used by Robo-advisors, have a greater possibility of earning unexpected returns, given its level of risk as measured by the Fuzzy Jensen's Alpha coefficient, than traditional investment strategies that use valuation for decision-making.*

# 3 Methodology

## 3.1 Data and Sample

To create a new portfolio strategy proposal for an RA platform as well as a value investing portfolio, we selected Latin American listed companies in Mexico, Colombia, Peru, Chile, and Brazil for the sample. Additionally, we used ETFs to replicate those countries. The timespan considered was from January 15, 2015, to June 28, 2019. It is worth mentioning that all the ETFs used for the research are from Blackrock.

First, to replicate the Brazilian market, we applied the EWZ. The EWZ's investment philosophy is to present exposure to large and medium-sized companies in Brazil. The ETF that we have used to replicate the movement of Peru is the EPU (iShares MSCI Peru ETF), knowing that the investment philosophy of replicating the results of the composite index of Peruvian equities. The key point of this ETF is that its total value is USD 143,014,932 (December 3, 2020), using the MSCI All Peru Capped Index as a reference index.

The ETF that we have used for Chile is the iShares MSCI Chile, also known as the ECH, whose investment philosophy consists of replicating the results of a broad-based index made up of Chilean equities that has a value of USD 509,355,508 of net assets, considering the MSCI Chile IMI 25/50 as a reference index. The ETF we have chosen for Colombia is the iShares Colombian MSCI, better known as ICOL. The investment philosophy consists of replicating the investment results of a broad-based index composed of shares of Colombian companies. It is valued at USD 21,270,772 of net assets (December 3, 2020) using the MSCI All Colombia Capped Index as a benchmark.

Finally, the ETF that we have used for Mexico is the iShares NAFTRAC, also known simply as NAFTRAC. This name is the result of a combination of the words of the initial creator Nacional Financiera and Tracker, being the first and most important ETF that replicates Mexico currently. Its investment philosophy consists of seeking results that correspond to the S&P/BMV IPC index.

For the first portfolio, a multiobjective function is used to optimize the risk-return ratio through genetic algorithms with EVOLVER, i.e., maximize the Sharpe ratio. For the second portfolio, a model is structured considering the recommendations of a consensus of market analysts from the Bloomberg platform. To assess the investment portfolios, the Fuzzy Jensen's Alpha is estimated to calculate the possibility of obtaining abnormal returns, similar to Cortez et al. [23], who applied the fuzzy regression method proposed by Tanaka [24] using crisp data from both portfolios.

To optimize the portfolios, we considered the GA approach [25]. A GA is an iterative procedure that maintains a constant size population $P(t)$ of candidate solutions. During each iteration step, called "a generation", the structures in the current population are evaluated, and based on those evaluations, a new population of candidate solutions is formed. The initial population $P(0)$ can be chosen heuristically or at random. The structures of the population $P(t+1)$ are chosen from $P(t)$ by a randomized selection procedure that ensures that the expected number of times a structure is chosen is approximately proportional to that structure's performance relative to the rest of the population. To search other points in the search space, some variation is introduced into the new population by means of idealized genetic recombination operators.

### 3.2 Fuzzy Alpha

A variety of techniques to evaluate the portfolio have been used, such as the Sharpe ratio, which is a risk-adjusted portfolio performance [26], and Jensen's alpha, which uses beta as the proper measure of risk [27]. In this research study, we will focus on the latter.

Jensen's alpha is a measure of abnormal performance [28]. The main purpose is to compare the performance of a certain portfolio (or index) to that of a benchmark portfolio [29], which is expressed in the following equation:

$$\alpha = R^p - \left[R^f + \beta\left(R^m - R^f\right)\right] \tag{2}$$

$$R^p = \sum_{k=1}^{s} x_k R_k$$

where:
$\alpha$ = Jensen's alpha.
$\beta$ = systemic risk.

$R^p$ = portfolio's expect return.
$R^f$ = return on risk-free rate.
$R^m$ = market portfolio's return.
$s$ = number of shares.
$R_k$ = expected return of the $k^{\text{th}}$ share.
$x_k$ = weight of the $k^{\text{th}}$ share in the portfolio.

Thus, to evaluate portfolio performance, Jensen's alpha was estimated since, as we mentioned above, the financial crisis is presented in the timespan considered; for this reason, a methodology contemplating uncertainty is needed. Cortez et al. [23] suggested the estimation of Jensen´s alpha with a fuzzy regression method, especially for uncertain periods.

Fuzzy regression models, similar to other regression techniques, seek to determine functional relationships between a dependent variable and one or more independent variables in which the $m$ parameters to be estimated are measured in confidence intervals through their centers and radii as follows:

$$A_i = A_{ic}, A_{ir} \tag{3}$$

$$A_{ic} = \frac{A_i^{max} + A_i^{min}}{2}; A_{ir} = \frac{A_i^{max} - A_i^{min}}{2}; i = 1, 2, \ldots, m$$

where:
$A_i^{min}$ = minimum of the confidence interval for the $i^{\text{th}}$ parameter.
$A_i^{max}$ = maximum of the confidence interval for the $i^{\text{th}}$ parameter.
$A_{ic}$ = enter of the confidence interval for the $i^{\text{th}}$ parameter.
$A_{ir}$ = radius of the confidence interval for the $i^{\text{th}}$ parameter.

We further assume that the relationship between the dependent and independent variables is linear. The objective of this process is to determine the centers and radii for these parameters that are consistent with the available observations. The goodness of fit is the inverse of the uncertainty (amplitude) of estimates of the observations. Therefore, the total uncertainty within the sample is the sum of the radii of the estimates:

$$z = \sum_{t=1}^{n} \hat{Y}_{rt} \tag{4}$$

$$\hat{Y}_{rt} = \sum_{i=0}^{p} A_{ir} |X_{it}|; \quad t = 1, 2, \ldots, n$$

where:
$z$ = total uncertainty.
$\hat{Y}_{rt}$ = amplitude estimated of the dependent variable at time $t$.
$A_{ir}$ = radius of the confidence interval for the $i^{\text{th}}$ parameter.
$X_{it}$ = $i^{\text{th}}$ independent variable at time $t$.
$p$ = total number of parameters to estimate.
$n$ = total number of observations.

Thus, the final goal is to minimize the total uncertainty of the estimates. The parameters should ensure not only that the uncertainty is as small as possible but also that estimates are as consistent as possible with the approximate observations of the dependent variable. To estimate the parameters, we applied the method proposed by Tanaka and Ishibuchi [24]. This method assumes that each observation ($Y_t$) must be included in the estimate $\left(\hat{Y}_t\right)$, i.e., $Y_t \subseteq \hat{Y}_t$, in accordance with the following conditions:

$$Y_{ct} - Y_{rt} \geq \hat{Y}_{ct} - \hat{Y}_{rt} \text{ and } Y_{ct} + Y_{rt} \leq \hat{Y}_{ct} + \hat{Y}_{rt} \tag{5}$$

Additionally, we employed crisp values that can also be used in the fuzzy regression method. In this sense, crisp values are those for which the extremes of a confidence interval are identical. If the central values are the crisp values with the radius set to zero, i.e., $Y_{rt} = 0$, then following Cortez et al. [23], the parameters of the fuzzy regression with crisp data are estimated by solving the next linear programming problem:

$$Min\ z = \sum_{t=1}^{n} \hat{Y}_{rt} = \sum_{t=1}^{n} \sum_{i=0}^{p} A_{ir}|X_{it}| \tag{6}$$

subject to:

$$Y_{ct} \geq \hat{Y}_{ct} - \hat{Y}_{rt} = \sum_{i=0}^{p} A_{ic}X_{it} - \sum_{i=0}^{p} A_{ir}|X_{it}|$$

$$Y_{ct} \leq \hat{Y}_{ct} + \hat{Y}_{rt} = \sum_{i=0}^{p} A_{ic}X_{it} + \sum_{i=0}^{p} A_{ir}|X_{it}|$$

$$A_{ir} \geq 0; \quad t = 1, 2, \ldots, n; \quad i = 0, 1, \ldots, p$$

Consequently, the fuzzy Jensen's alpha with crisp observations is estimated with the fuzzy regression method of Eq. (7) considering the following model:

$$\left\langle \hat{R}^p_{ct}, \hat{R}^p_{rt} \right\rangle = \langle \alpha_c, \alpha_r \rangle + \langle \beta_c, \beta_r \rangle R^m_t \tag{7}$$

$$t = 1, 2, \ldots, n$$

where:
$\langle \alpha_c, \alpha_r \rangle$ = fuzzy Jensen's alpha indicated by its center and radius.
$\langle \beta_c, \beta_r \rangle$ = fuzzy systemic risk.
$\left\langle \hat{R}^p_{cj}, \hat{R}^p_{rj} \right\rangle$ = estimated fuzzy portfolio return at time $t$.
$R^m_t$ = market portfolio's return at time $t$.
$n$ = number of observations.
The next step is to find the shares' weight combination within the portfolio that maximizes the Sharpe ratio. Additionally, to ensure that the portfolio includes all the

ETFs of the sample, a restriction was added. In this sense, the nonlinear programming problem is stated as follows:

$$Max \frac{R_p - R_f}{\sigma_p}, \quad (8)$$

subject to:

$$0.01 \leq x_k \leq 1; \sum_{k=1}^{s} x_k = 1$$

$$t = 1, 2, \ldots, n$$

where:
$R_p$ is the return of the portfolio.
$R_f$ is the risk-free rate; and.
$\sigma_p$ is the standard deviation of the portfolio.
$x_k$ = weight of the $k^{th}$ share in the portfolio.
$s$ = number of shares.
$n$ = number of observations.

### 3.3 Design of the Portfolio

We designed a semiautomatic robot through Excel and visual basic, where the strategy was to take the latest historical data from the last 20 days of the assets included in the portfolio, as shown in Fig. 3.

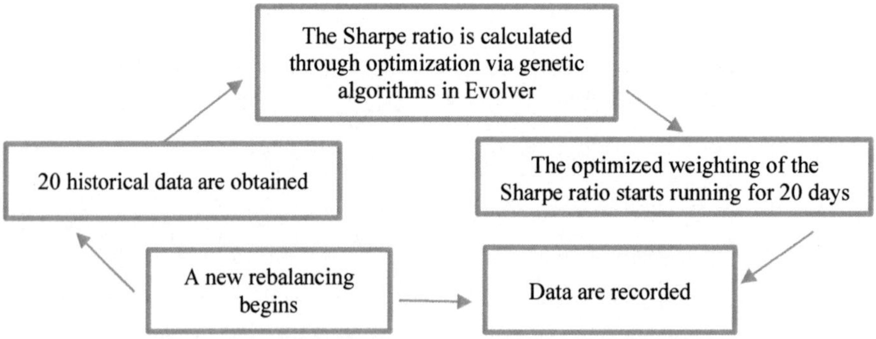


**Fig. 3.** Design of the Robo Advisor model. Source: Own elaboration.

With this historical base, the Sharpe ratio is calculated, which is maximized by means of genetic algorithms through a technological tool called Evolver. Rebalancing is performed by the algorithm every 20 days to obtain the new weights. Next, we present a small model explaining how the Robo advisor works.

The value investing portfolio, was built using the consensus of analysts who followed up on each of the companies in the sample of the five Latin American countries. Consequently, two restrictions were established to ensure the filtering of the best stations and include them in the portfolio. The first restriction is that at least 4 analysts must follow the company, and the second restriction is that at least 70% of analysts recommend purchasing it. Likewise, the RA portfolio maintains a 20-day rebalancing where the best company share that can participate in the portfolio is reevaluated based on the two restrictions. It is important to mention that once the two restrictions are overcome, all actions that pass this filter are weighted equally. Figure 4 shows the design of the value investing analyst model.

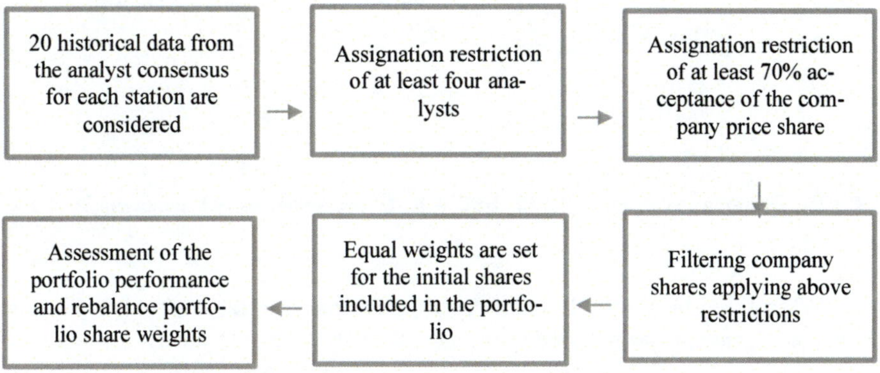


**Fig. 4.** Design of the value investing analyst model. Source: Own elaboration.

For the benchmarking, we applied an equitable weighting for the ETFs of the 5 countries that we analyzed in the study (Chile, Mexico, Brazil, Peru, and Colombia). The design of the benchmark arises from the need to evaluate the strategies in a neutral position, which remains lacking in a strategist but remains positioned within all assets that have both the theft advisor strategy and the value investing analyst strategy.

Finally, to compare the results between the RA and the value investing portfolios, we consider the possibility of obtaining the abnormal returns metric proposed by Cortez and Rodriguez [23] as follows:

$$Poss^{AR}(R^p) = \begin{cases} 1 & \text{if } \alpha^{min} > 0, \alpha^{max} > 0 \\ \frac{\alpha^{max}}{|\alpha^{min}| + \alpha^{max}} & \text{if } \alpha^{min} \langle 0, \alpha^{max} \rangle 0 \\ 0 & \text{otherwise} \end{cases} \tag{9}$$

where:
$Poss^{AR}(R^p)$ = possibility that the portfolio obtains abnormal returns.
$\alpha^{min}$ = minimum value of the fuzzy Jensen's alpha.
$\alpha^{max}$ = maximum value of the fuzzy Jensen's alpha.

## 4 Results

Figure 5 shows the performance for both strategies. From 2015 to 2018, the evolution of both series is quite similar; however, after 2018, the value investing strategy outperforms both RA and Benchmark, creating a gap between them.

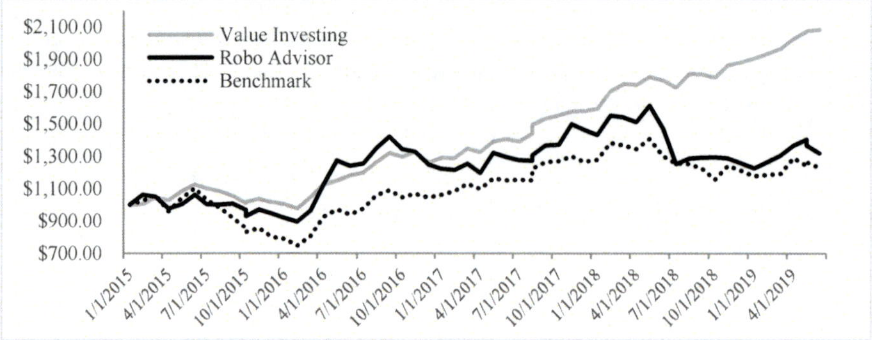


**Fig. 5.** Temporary investment of 1000 USD with the two strategies vs benchmark. Source: Own elaboration.

When observing the results of the descriptive statistics in Table 1, the RA has a higher standard deviation and standard error than the analysts do. Originally, it was expected that the RA would have less volatility derived from greater diversification; however, it can be observed that the Sharpe ratio optimization strategy via genetic algorithms overweights the portfolio to one or two ETFs in most rebalances. This effect causes greater concentration and, therefore, greater volatility. We also observe that the gap between the low and high values of the RA is larger than those of the analysts' portfolio, derived from the same overweight effect when rebalancing every twenty days.

**Table 1.** Descriptive statistics

| Statistic | Analysts | Robo advisor | Benchmark |
|---|---|---|---|
| Mean | 1.4% | 0.6% | 0.4% |
| Median | 1.3% | −0.5% | 0.0% |
| Geometric Mean | 1.31% | 0.44% | 0.27% |
| Standard Error | 0.4% | 0.7% | 0.7% |
| Standard Deviation | 2.8% | 5.6% | 5.1% |
| Sample Variance | 0.1% | 0.3% | 0.3% |
| Range | 12% | 31% | 25% |
| Minimum | −5.0% | −14.9% | −9.6% |

*(continued)*

**Table 1.** (*continued*)

| Statistic | Analysts | Robo advisor | Benchmark |
|---|---|---|---|
| Maximum | 7.3% | 16.5% | 15.0% |
| Semi deviation | 1.91% | 3.43% | 3.21% |
| Standarized Skewness | 0.0318 | 1.1799 | 1.2097 |
| Standarized Kurtosis | −0.7657 | 1.4896 | −0.1130 |
| Jarque Bera (JB) | 0.0638 | 1.6298 | 1.4646 |
| JB p-value | (0.9686) | (0.4427) | (0.4808) |
| Count | 56 | 56 | 56 |

Nevertheless, when comparing both strategies using the possibility of obtaining abnormal returns, the results are different. In this case, the RA portfolio has a 62% possibility of obtaining unexpected returns (see Fig. 6).

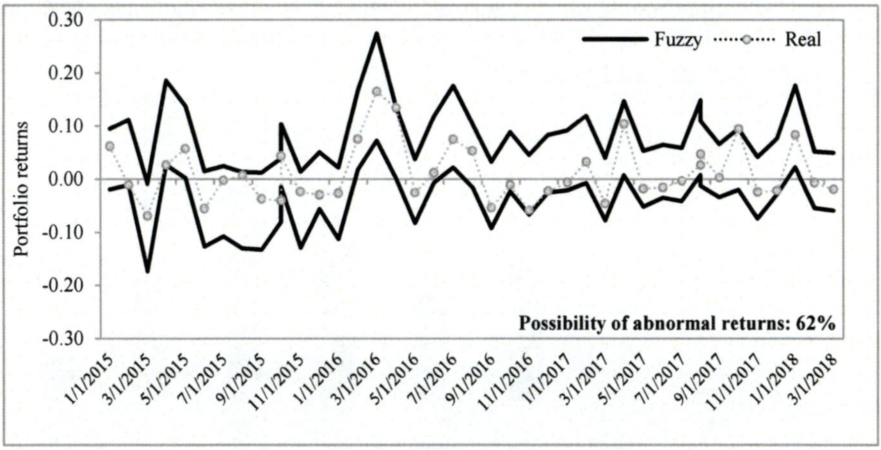


**Fig. 6.** Fuzzy regression for the RA portfolio.

On the other hand, the possibility of abnormal returns of the value investing portfolio is 49% (see Fig. 7). Therefore, the RA portfolio offers us a greater possibility of obtaining unexpected returns than the value investing portfolio.

As we can see in the Fig. 8, the portfolio prepared by the analysts gives an average return higher than the average return of the RA; however, there is a greater possibility of abnormal, unexpected returns in the case of the RA. If we examine the distribution of the data, we can provide some explanations for these results.

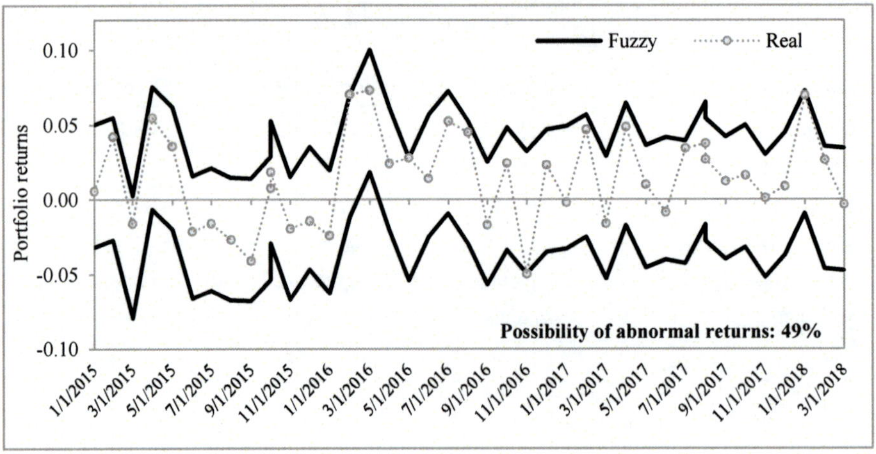


**Fig. 7.** Fuzzy regression for the value investing portfolio. Source: Own elaboration.

Figure 8 shows the distribution of the returns for the value investing strategy. The returns are mainly concentrated at approximately 3.79% with a frequency of 14 times, followed by returns over 2.03% with a frequency of 11 times. Alternatively, negative variations were presented 11 times.

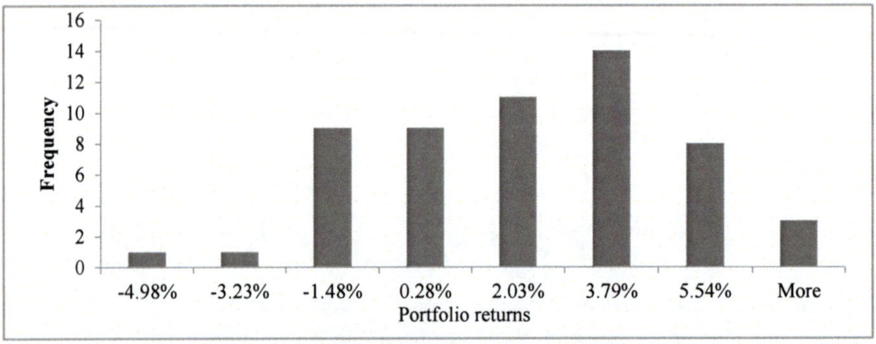


**Fig. 8.** Histogram for the value investing strategy. Source: Own elaboration.

On the other hand, Fig. 9 shows the histogram for the RA strategy. We observe a greater fluctuation in returns compared to the value investing strategy, varying from −14% to 12.02%. The highest frequency was located at approximately −1.42%, followed by 3.06% with a frequency of 16 times and 7.54% with a frequency of 11 times. Thus, we can see that the returns with greater frequency are negative; however, the positive sessions with greater frequency (7.54%, Fig. 9) are significantly higher than those of the value investing portfolio (3.79%, Fig. 8).

Finally, analyzing the benchmarking in Fig. 10, the highest values in the histogram are at −3% and 4%, sharing a frequency of 14 times. The total frequency of negative

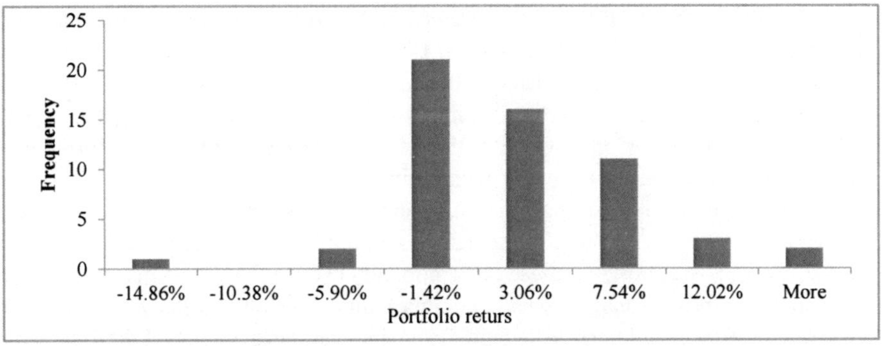


**Fig. 9.** Histogram for the RA strategy. Source: Own elaboration.

returns is 20 times, placing the most negative return at −10% and the most positive return at 12.02%.

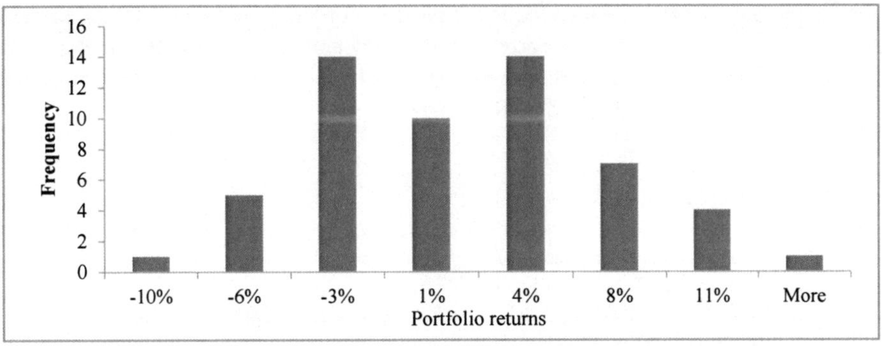


**Fig. 10.** Histogram for the benchmark. Source: Own elaboration.

## 5 Conclusions

After observing the results, we can report that both strategies succeeded in surpassing the benchmark; however, the value investing portfolio accelerated its growth after 2018, increasing its positive gap against the RA portfolio. In this sense, the portfolio developed by the analysts gives an average return higher than the average return of the RA. Nonetheless, the RA portfolio has a greater possibility of obtaining abnormal or unexpected returns than the analyst's value investing portfolio, given the systematic risk involved.

Despite the acceptance and exponential growth of RA platforms in the United States over the last ten years, emerging markets still maintain a degree of backwardness, probably due to the cultural shock with technology in these countries. Although it is kwon, there are currently a couple of attempts of this type of platform; at this moment, there is no RA that presents all the characteristics in the portfolio management area to offer competitive administrative commissions. For this reason, it is considered relevant to continue carrying out studies related to the construction of automated portfolios for RA platforms to better understand the advantages and disadvantages of this type of structure and to seek continuous improvement. Furthermore, promoting those platforms would lead to their more rapid acceptance in emerging countries.

On the other hand, the impact that this study has on stock market finance is relevant since it presents more evidence of the behavior of portfolios with methodologies such as value investing and similar to those conventionally used in RA platforms. This evidence will allow portfolio managers to add another perspective on investor behavior and the risk-return relationship of these techniques inside emerging markets. Likewise, we can conclude that each portfolio offers different characteristics, and understanding them will provide guidelines for better portfolio management.

Thus, this study concludes that the value investing methodology may not be substitutable; however, methodologies similar to those of RA promise greater possibilities of abnormal returns. Finally, regarding new lines of research, this study expects to continue looking forward to new ways of optimizing management strategies of RA platforms since the Sharpe ratio tends to focus the portfolio on only one or two assets; therefore, future research could explore correcting this concentration. Likewise, there are many other techniques that could be combined to find a better risk-to-return ratio using passive management, such as that used by RA platforms.

## References

1. P Gomber JA Koch M Siering 2017 Digital finance and FinTech: current research and future research directions J. Bus. Econ. 87 5 537 580
2. Fein, M.L.: Robo-Advisors: A closers look. SSRN Working Paper. https://ssrn.com/abstract=2658701 (2015)
3. ICI: Investment Company Fact Book, 58th edn. Investment Company Institute, Washington, DC (2018)
4. M Beketov K Lehmann M Wittke 2018 Robo advisors: quantitative methods inside the robots J. Asset Manag. 19 6 363 370
5. Melone, C.: Investors Attitudes Towards Robo-advisors. My Private Banking Research, Kreuzlingen, Switzerland (2016)
6. L Bjerknes A Vukovic 2017 Automated advice: A Portfolio Management Perspective on Robo-advisors Norwegian University of Science and Technology, Trondheim Master's thesis
7. MW Uhl P Rohner 2018 The compensation portfolio Financ. Res. Lett. 27 60 64
8. A Ling J Sun M Wang 2019 Robust multi-period portfolio selection based on downside risk with asymmetrically distributed uncertainty set Eur. J. Oper. Res. 285 1 81 95
9. R Bird J Whitaker 2004 The performance of value and momentum investment portfolios: Recent experience in the major European markets part 2 J. Asset Manag. 5 3 157 175
10. F D'Acunto N Prabhala AG Rossi 2019 The promises and pitfalls of robo-advising Oxford Univ. Press behalf of The Society for Financial Studies 32 5 1983 2020
11. B Graham DL Dodd 2009 Security Analysis 6 McGraw Hill New York
12. EF Fama KR French 1993 Common risk factors in the returns on stocks and bonds J. Financ. Econ. 33 3 56
13. A Damodaran 2005 Value and risk: beyond betas Financ. Analyst J. 61 2 38 43

14. C Asness A Frazzini R Israel T Moskowitz 2015 Fact, fiction, and value investing J. Portfolio Manage. 42 1 34 52
15. SD Campbell SA Sharpe 2009 Anchoring bias in consensus forecasts and its effect on market prices J. Financ. Quant. Anal. 44 2 369 390
16. E Otuteye M Siddiquee 2015 Overcoming cognitive biases: heuristic for making value investing decisions J. Behav. Financ. 16 2 140 149
17. H Markowitz 1952 Porfolio selection J. Financ. 7 1 7 91
18. T Bodnar T Zabolotskyy 2016 How risky is the optimal portfolio which maximizes the Sharpe ratio? Adv. Stat. Anal. 101 1 1 28 https://doi.org/10.1007/s10182-016-0270-3
19. WF Sharpe 1966 Mutual fund performance J. Bus. 39 1 119 138
20. Sironi, P.: Fintech innovation: from robo-advisors to goal based investing and gamification, 1st edn. Wiley Finances Series, Hoboken (2016)
21. K Phoon F Koh 2018 Robo-advisors and wealth management Alternat. Invest. 20 3 79 94
22. O Ivanov O Snihovyi V Kobets 2018 Implementation of robo-advisors tools for different risk attitude investment decisions CEUR Workshop Proceedings 2104 161 195 206
23. K Cortez M Rodríguez B Mendez 2013 An Assessment of abnormal returns and risk in socially responsible firms using fuzzy alpha Jensen and fuzzy beta Fuzzy Econ. Rev. 18 1 37 59
24. Tanaka, H., Ishibuchi, H.: Possibilistic regression analysis based on linear programming. In: Kacprzyk, J., Fedrizzi M. (eds.) Fuzzy Regression Analysis, pp. 47–60. Omnitech Press, Warsaw and Physica-Verlag, Heidelberg (1992)
25. J Grefenstette 1986 Optimization of control parameters for genetic algorithms IEEE Trans. Syst. Man Cybern. 16 1 122 128
26. JE Grable S Chatterjee 2014 Reducing wealth volatility: the value of financial advice measured by zeta J. Financ. Plan. 27 8 45 51
27. CM Jensen 1968 The performance of mutual funds in the period 1945–1964 J. Financ. 23 2 389 416
28. R Jarrow P Protter 2013 Positive alphas, abnormal performance, and illusory arbitrage Math. Financ. 23 1 39 56
29. S Flaherty J Li 2004 Composite performance measures Chin. Econ. 37 3 39 66

# Author Index





M. d. P. Rodríguez García et al. (Eds.): XX SIGEF 2021, LNNS 384, p. 195, 2022.
https://doi.org/10.1007/978-3-030-94485-8

14. C Asness A Frazzini R Israel T Moskowitz 2015 Fact, fiction, and value investing J. Portfolio Manage. 42 1 34 52
15. SD Campbell SA Sharpe 2009 Anchoring bias in consensus forecasts and its effect on market prices J. Financ. Quant. Anal. 44 2 369 390
16. E Otuteye M Siddiquee 2015 Overcoming cognitive biases: heuristic for making value investing decisions J. Behav. Financ. 16 2 140 149
17. H Markowitz 1952 Porfolio selection J. Financ. 7 1 7 91
18. T Bodnar T Zabolotskyy 2016 How risky is the optimal portfolio which maximizes the Sharpe ratio? Adv. Stat. Anal. 101 1 1 28 https://doi.org/10.1007/s10182-016-0270-3
19. WF Sharpe 1966 Mutual fund performance J. Bus. 39 1 119 138
20. Sironi, P.: Fintech innovation: from robo-advisors to goal based investing and gamification, 1st edn. Wiley Finances Series, Hoboken (2016)
21. K Phoon F Koh 2018 Robo-advisors and wealth management Alternat. Invest. 20 3 79 94
22. O Ivanov O Snihovyi V Kobets 2018 Implementation of robo-advisors tools for different risk attitude investment decisions CEUR Workshop Proceedings 2104 161 195 206
23. K Cortez M Rodríguez B Mendez 2013 An Assessment of abnormal returns and risk in socially responsible firms using fuzzy alpha Jensen and fuzzy beta Fuzzy Econ. Rev. 18 1 37 59
24. Tanaka, H., Ishibuchi, H.: Possibilistic regression analysis based on linear programming. In: Kacprzyk, J., Fedrizzi M. (eds.) Fuzzy Regression Analysis, pp. 47–60. Omnitech Press, Warsaw and Physica-Verlag, Heidelberg (1992)
25. J Grefenstette 1986 Optimization of control parameters for genetic algorithms IEEE Trans. Syst. Man Cybern. 16 1 122 128
26. JE Grable S Chatterjee 2014 Reducing wealth volatility: the value of financial advice measured by zeta J. Financ. Plan. 27 8 45 51
27. CM Jensen 1968 The performance of mutual funds in the period 1945–1964 J. Financ. 23 2 389 416
28. R Jarrow P Protter 2013 Positive alphas, abnormal performance, and illusory arbitrage Math. Financ. 23 1 39 56
29. S Flaherty J Li 2004 Composite performance measures Chin. Econ. 37 3 39 66

# Author Index




M. d. P. Rodríguez García et al. (Eds.): XX SIGEF 2021, LNNS 384, p. 195, 2022.
https://doi.org/10.1007/978-3-030-94485-8